A Practical Guide to
Cancer Systems Biology

A Practical Guide to Cancer Systems Biology

Edited by

Hsueh-Fen Juan
National Taiwan University, Taiwan

Hsuan-Cheng Huang
National Yang-Ming University, Taiwan

World Scientific

NEW JERSEY · LONDON · SINGAPORE · BEIJING · SHANGHAI · HONG KONG · TAIPEI · CHENNAI · TOKYO

Published by

World Scientific Publishing Co. Pte. Ltd.

5 Toh Tuck Link, Singapore 596224

USA office: 27 Warren Street, Suite 401-402, Hackensack, NJ 07601

UK office: 57 Shelton Street, Covent Garden, London WC2H 9HE

Library of Congress Cataloging-in-Publication Data

Names: Juan, Hsueh-Fen, author. | Huang, Hsuan-Cheng, author.

Title: A practical guide to cancer systems biology / Hsueh-Fen Juan, National Taiwan University, Taiwan,
    Hsuan-Cheng Huang, National Yang-Ming University, Taiwan.

Description: New Jersey : World Scientific, [2017] | Includes bibliographical references and index.

Identifiers: LCCN 2017026929 | ISBN 9789813229143 (hardcover : alk. paper)

Subjects: LCSH: Cancer--Research--Data processing. | Cancer--Research--Methodology. | Systems biology.

Classification: LCC RC267 .J83 2017 | DDC 616.99/40072--dc23

LC record available at https://lccn.loc.gov/2017026929

British Library Cataloguing-in-Publication Data

A catalogue record for this book is available from the British Library.

For any available supplementary material, please visit
http://www.worldscientific.com/worldscibooks/10.1142/10685#t=suppl

Typeset by Stallion Press
Email: enquiries@stallionpress.com

Contents

Chapter 1   Introduction to Cancer Systems Biology   1
             Hsueh-Fen Juan and Hsuan-Cheng Huang

Chapter 2   Transcriptome Analysis: Library Construction   11
             Hsin-Yi Chang and Hsueh-Fen Juan

Chapter 3   Quantitative Proteome: The Isobaric Tags for
             Relative and Absolute Quantitation (iTRAQ)   23
             Yi-Hsuan Wu and Hsueh-Fen Juan

Chapter 4   Phosphoproteome: Sample Preparation   39
             Chia-Wei Hu and Hsueh-Fen Juan

Chapter 5   Transcriptomic Data Analysis: RNA-Seq Analysis
             Using Galaxy   49
             Chia-Lang Hsu and Chantal Hoi Yin Cheung

Chapter 6   Proteomic Data Analysis: Functional Enrichment   63
             Hsin-Yi Chang and Hsueh-Fen Juan

Chapter 7   Phosphorylation Data Analysis   75
             Chia-Lang Hsu and Wei-Hsuan Wang

Chapter 8   Pathway and Network Analysis   91
             Chen-Tsung Huang and Hsueh-Fen Juan

Chapter 9   Dynamic Modeling   103
             Yu-Chao Wang

Chapter 10   Protein Structure Modeling   113
              Chia-Hsien Lee and Hsueh-Fen Juan

Chapter 11   Docking Simulation   129
              Chia-Hsien Lee and Hsueh-Fen Juan

Index   145

1. Introduction to Cancer Systems Biology

Hsueh-Fen Juan[*,‡] and Hsuan-Cheng Huang[†]

*Department of Life Science, Institute of Molecular and Cellular Biology,
and Graduate Institute of Biomedical Electronics and Bioinformatics,
National Taiwan University, Taipei 106, Taiwan
†Institute of Biomedical Informatics, Center for Systems and Synthetic Biology,
National Yang-Ming University, Taipei 112, Taiwan
‡yukijuan@ntu.edu.tw

1. Cancer systems biology

Cancer is a life-threatening disease and it is one of the leading causes of death worldwide.[1] Our knowledge of cancer is advancing at an astonishing rate; however, the more we understand, the more complexities cancer reveals.[2] Various processes, at both the molecular and cellular levels, function in a dynamic and connective manner which must be understood to address the disease.[2] Probing global DNA, RNA, and protein expression can generate big omics data which in turn creates a systems approach to cancer biology research, also known as cancer systems biology. Cancer systems biology addresses the complexity of cancer by integrating experimental and computational methods in the synthesis and testing of cancer biological hypotheses.

2. Omics big data and bioinformatics

Systems biology is a data-driven science which produces big omics data for the discovery of disease biomarkers and drugs, as well as the study of molecular regulation and mechanism (Fig. 1). After the human genome profile is completed, the post-genome era would be upon us. The biological omics data include genome, transcriptome, proteome, and metabonome. How

1

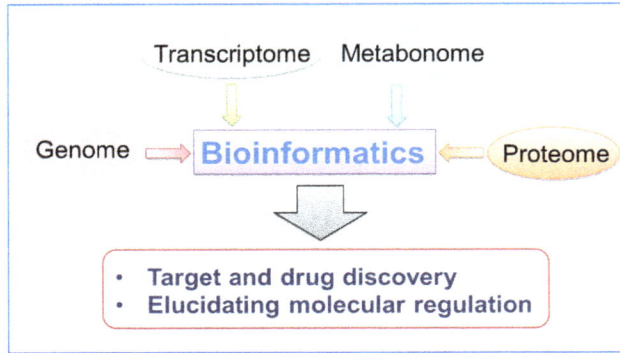

Figure 1. The relationship between bioinformatics and omics data. Systems biology is the first to generate biological big data — such as genome, transcriptome, proteome, and metabonome — and then analyzes the data using a bioinformatics approach. In this way, we can discover disease biomarkers and drugs as well as elucidate molecular regulations in living organisms.

can we find important information in a sea of biological data? This is a challenge that bioinformatics can tackle.

2.1. *Genome*

A genome is a complete set of genes or genetic material in a cell or organism. The study of genomes is known as "genomics", and it aims to explore the DNA content and structure as well as the function and evolution of genomes. From the data in National Center for Biotechnology Information (NCBI, https://www.ncbi.nlm.nih.gov/genome/browse/), 4,175 eukaryote and 95,217 prokaryote genomes have been sequenced. The size of the human genome is over three billion base pairs which are organized into 22 paired chromosomes and the XX or XY chromosome. The human genome contains coding and non-coding DNA. Coding DNA can be transcribed into mRNAs and further translate into proteins; however, non-coding DNA can be transcribed into RNAs but cannot be encoded into proteins.

The Cancer Genome Atlas (TCGA), previously known as the Human Cancer Genome Project, has generated comprehensive, multidimensional maps of the key genomic changes in 33 different tumor types in 11,000 patients (https://cancergenome.nih.gov/). TCGA dataset has a total of 2.5 PB of genomic data which were generated not only by microarrays but also by using next generation sequencing techniques. These data are publicly available.

2.2. *Transcriptome*

The central dogma is an explanation of the flow of genetic information from DNA to RNA to protein in a biological system. A transcriptome is a set of mRNAs or total set of transcripts in a cell or organism. We also call the transcriptome the gene expression profile because it examines the expression level of mRNAs. The study of transcriptomes is termed transcriptomics. The high-throughput techniques on transcriptome include DNA microarray and next-generation sequencing (RNA-sequencing).

TCGA dataset includes not only genomics data but also transcriptomics and proteomics data. The Gene Expression Omnibus (GEO) is a public functional genomics data repository which accepts array- and sequence-based data (https://www.ncbi.nlm.nih.gov/geo/). ArrayExpress is the other functional genomics data archive supported by the European Bioinformatics Institute (EMBL-EBI, https://www.ebi.ac.uk/arrayexpress/). TCGA, GEO, and ArrayExpress provide their data for reuse to the research community.

2.3. *Proteome*

The definition of proteome is the protein set in a cell, a tissue and an organism at a certain time. The study of proteomes is known as proteomics. The term "proteome" was coined by Marc Wilkins in 1994. The most popular and high-throughput technique in proteomics is liquid chromatography–mass spectrometry (LC–MS). Proteomics can measure protein expression level and discover post-translational modifications such as phosphorylation, ubiquitination and acetylation. Post-translational modifications affect the activities of proteins to regulate cellular progression and development.

The ProteomeXchange (PX) Consortium (http://www.proteomex change.org) was founded in 2011 to provide proteomics resources and standardize data submission and dissemination of mass spectrometry (MS)-based proteomics data.[3] PX contains 4,534 datasets. The top five species are: *Homo sapiens, Mus musculus, Saccharomyces cerevisiae, Arabidopsis thaliana* and *Rattus norvegicus*. Since 2015, the journal *Molecular and Cellular Proteomics*, one of the most prominent scientific proteomics journals, has mandated the deposition of raw data with every submitted paper.

2.4. *Metabolome*

The metabolome is defined as the complete set of small-molecule chemicals in a cell, a tissue or an organism. Metabolomics is the study of small molecules, such as sugars and amino acids, which are substrates, products, or intermediates of metabolic reactions in an organism. The most used methods to measure metabolome are nuclear magnetic resonance (NMR) spectroscopy and MS. One of the most popular metabolomics database is Kyoto Encyclopedia of Genes and Genomes (KEGG, http://www.genome.jp/kegg/).

2.5. *Bioinformatics*

Omics techniques generate big data such as genome, transcriptome, proteome, and metabolome. How to integrate and mine these data is an important issue.[4] There is no doubt that bioinformatics or computational biology can help researchers analyze these big omics data. The categories of bioinformatics contain many subfields such as sequence analysis, genome annotation, transcriptome and proteome analyses, as well as network biology and dynamic modeling.[5]

Both network biology and modeling are important in the study of systems biology.[1] Network biology includes network and pathway analyses which help researchers understand the relationships within metabolic, protein–protein interaction networks and phospho-signaling pathways. Dynamic modeling requires gene or protein expression time-series data and mathematical methods (Fig. 2).[5] Docking simulation such as protein-drug or DNA-drug docking has the potential to enhance drug development.[6] These methods can be beneficial for disease biomarker and target identification as well as drug discovery.

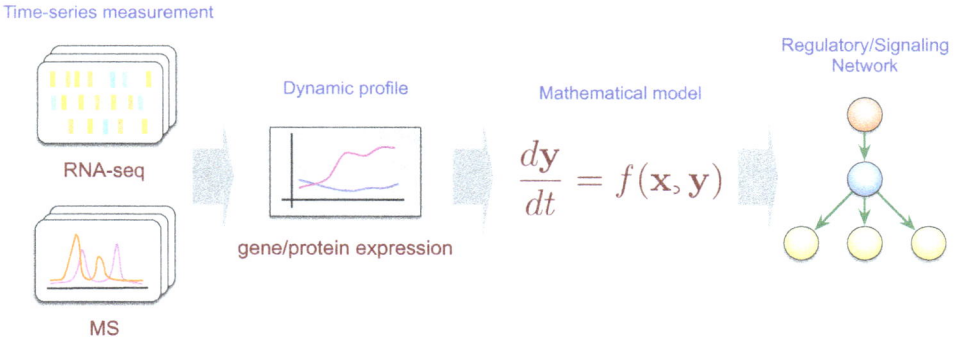

Figure 2. Dynamic modeling. Gene or protein expression time-series data and mathematical methods are required to infer regulatory or signaling networks.

3. Systems biology in cancer research

Cancer is a complex disease. The cancer systems biology approach is very effective in identifying drug targets, developing novel therapeutics and new indications for existing drugs.[7–10] In addition to target and drug discovery, cancer systems biology represents a valuable contribution to efforts toward understanding the molecular mechanism of drugs in various cancer types[5,11] and regulation in cancer progression.[4]

3.1. *Target and drug discovery*

Omics data in systems biology encompasses genome, transcriptome, proteome, and metabolome. Reports using omics data in identifying therapeutic targets and drugs have exponentially increased in the past decade. Here we will introduce research work on target and drug discovery using genomics, transcriptomics and proteomics approaches as well as bioinformatics and network biology.

We can identify potential targets and biomarkers by mining TCGA data — in particular, genomics and transcriptomics data. The targets and biomarkers can be protein-coding mRNA and non-coding RNA such as long non-coding RNAs (lncRNA) and microRNAs. In our recent work, we identified 6 lncRNAs by analyzing the expression profiles of lncRNAs and protein-coding genes between MYCN amplified and MYCN non-amplified neuroblastoma (NB) patient samples from microarray and RNAseq datasets and performing network analysis.[10] In the study, we found SNHG1 could be a potential prognostic biomarker for high-risk NB intervention. Additionally, we identified several microRNAs, such as miR-124-3p and miR-93-5p, and RNA binding proteins: Synaptotagmin binding, cytoplasmic RNA-interacting protein (SYNCRIP) and insulin-like growth factor-2 mRNA-binding protein 3 (IGF2BP3), which could be potential therapeutic targets for NB.[4]

Besides genomics and transcriptomics, proteomics can also be used to discover potential targets and biomarkers.[12] In our previous studies, we utilized two dimensional (2D)-based proteomics coupling MS to identify therapeutic targets or biomarkers for breast cancer,[9] lung cancer,[13] and gastric cancer.[14,15] Using docking simulation for novel targets is helpful in revealing their inhibitors. In our previous study, ATP synthase is a potential drug target for breast and lung cancers. We further preformed homology modeling and docking simulation and found aurovertin B and citreoviridin to be potential drugs for breast and lung cancer.[8,13] In the final two chapters of this book, we will demonstrate how to use homology modeling and docking simulation for drug discovery.

3.2. *Molecular regulation*

3.2.1. *Gene expression profiles for molecular regulation*

Comparing the gene expression profiles after drug treatment or not, we are able to study the molecular mechanism of the drug in cancer cells. Combining time-series gene expression measurements and network construction, and further validation by real-time quantitative polymerase chain reaction (Q-PCR), scientists can reconstruct novel drug-response networks which provide molecular explanations of the drugs. Using this approach, we reconstructed an apoptosis-related gene network which is induced by a novel synthetic compound cRGD in human breast cancer cells.[16] This study may create an impact on breast cancer therapy.

This strategy of using gene expression and network studies can not only be applied to synthetic drugs but also to natural products. *Ganoderma lucidum* has been used in traditional Chinese medicine with antitumor and immuno-modulatory effects. From microarray gene expression studies and network construction, we found novel molecular mechanism in natural products. As shown in the papers, we extracted natural products from *Ganoderma lucidum* and found polysaccharides designated as F3 can induce macrophage-like differentiation via receptor oligomerization, recruitment of specialized adaptor proteins, caspase and p53 activation.[17,18]

The strategy described above can also be used to study molecular regulation. In these studies, we used proteomics to discover annexin A4 (ANXA4) as a novel biomarker for gastric cancer and further performed exon array to elucidate its molecular function in gastric cancer cells. We demonstrated that ANXA4 binds to plasma membrane in a Ca^{2+}-dependent manner and induces downstream signaling transduction that promote cell growth.[19]

3.2.2. *Proteomes for molecular regulation studies*

Proteomics can speed up the process of discovering cancer biomarkers and drug targets as well as the study of molecular regulations of drugs or non-coding RNAs. The global changes of protein expression in normal and disease situations can be visualized and compared with 2-DE and further identified by MS analysis.[8,14,15] ATP synthase was highly expressed in breast and lung cancer patients and reported as a potential drug target.[8,13] Additionally, using homology modeling and docking simulation, aurovertin B and citreoviridin were screened out to inhibit ATP synthase and found to trigger cancer cell death. We proved that citreoviridin-induced response network was involved in protein folding by proteomics and phosphoproteomics[5,13] Furthermore, targeting both ectopic ATP synthase and 26s proteasome induces ER stress in breast cancer cells.[20] *In vivo* study showed that citreoviridin induces alterations in the expression of glucose metabolism-related enzymes in lung cancer.[21]

The tripeptide sequence RGD (Arg-Gly-Asp) is a common cell recognition motif, which is a part of integrin-binding ligands.[22] We synthesized a cyclic RGD (Tpa-RGDWPC, cRGD) with rigid skeleton that can bind closely to its acceptor and further measured the expression profile of cRGD-treated MCF-7 cells with time-course proteomics, and finally used clustering analysis to reveal the temporal patterns of altered protein expression that can be classified as early, intermediate, and late response proteins.[22] These results demonstrate a molecular explanation for the properties of cRGD in cancer cells and provide valuable insight towards their role in cancer therapy.[22]

In addition to the study of molecular regulations of drugs, proteomics can also help to study the molecular mechanism of non-coding RNA such as microRNAs (miRNAs). MiRNAs are a class of small, endogenous, and highly conserved non-coding RNAs that are found abundantly in eukaryotic cells.[23] In our previous study, we combined a quantitative proteomics and stable isotope labeling to identify the global profile of miR-148a-regulated downstream proteins which were involved in neural development. Further validation by silencing of miR-148a in zebrafish found the abnormal morphology and decreased expression of neuron-related markers in the developing brain.[24] This convincing example shows that proteomics can facilitate the discovery of novel functions in non-coding RNAs.

Integrating transcriptomics and proteomics is useful in revealing drug response mechanisms. For example, tanshinone IIA (TIIA) is a diterpene quinone extracted from the plant Danshen (*Salvia miltiorrhiza*), which is used in traditional Chinese medicine.[11,25] By analyzing RNA-sequencing and proteome data in TIIA-treated gastric cancer cells, TIIA was found to suppress cell growth by blocking glucose metabolism.[11]

4. Conclusion

Systems biology requires omics techniques which generate high-throughput data and enable a holistic view of gene or protein expression related to cancer, hence bioinformatics is critical for data analysis. Elucidation of omics data from cancer patients and cells enhances our ability to identify biomarkers and drug targets as well as to study the molecular mechanisms of drugs. According to the targets, we can utilize bioinformatics to pick out potential anti-cancer drugs. Additionally, systems biology approaches will contribute in providing effective strategies for cancer therapy using both single drug and combinations.

In this book, we will provide the protocols needed to generate omics data, especially transcriptome and proteome, and tutorials that show readers how to analyze these data using bioinformatics as well as to discover drug candidates using docking simulation.

References

1. Juan HF and Huang HC. (2012) *Systems Biology: Applications in Cancer-related Research*, World Scientific Publishing, Singapore.
2. Gentles AJ and Gallahan D. (2011) Systems biology: Confronting the complexity of cancer. *Cancer Res.* **71:** 5961–5964.
3. Deutsch EW, Csordas A, Sun Z *et al.* (2017) The ProteomeXchange consortium in 2017: Supporting the cultural change in proteomics public data deposition. *Nucleic Acids Res.* **45:** D1100–D1106.
4. Hsu CL, Chang HY, Chang JY *et al.* (2016) Unveiling MYCN regulatory networks in neuroblastoma via integrative analysis of heterogeneous genomics data. *Oncotarget* **7:** 36293–36310.
5. Hu CW, Hsu CL, Wang YC *et al.* (2015) Temporal phosphoproteome dynamics induced by an ATP synthase inhibitor citreoviridin. *Mol. Cell Proteomics* **14:** 3284–3298.
6. Lee CH, Huang HC, and Juan HF. (2011) Reviewing ligand-based rational drug design: The search for an ATP synthase inhibitor. *Int. J. Mol. Sci.* **12:** 5304–5318.
7. Berg EL. (2014) Systems biology in drug discovery and development. *Drug Discov. Today* **19:** 113–125.
8. Huang TC, Chang HY, Hsu CH *et al.* (2008) Targeting therapy on breast carcinoma by ATP synthase inhibitor aurovertin B. *J. Proteome Res.* **7:** 1433–1444.
9. Sahu D, Hsu CL, Lin CC *et al.* (2016) Co-expression analysis identifies long noncoding RNA SNHG1 as a novel predictor for event-free survival in neuroblastoma. *Oncotarget* **9:** 58022–58037.
10. Lin LL, Huang HC, and Juan HF. (2014) Deciphering molecular determinants of chemotherapy in gastrointestinal malignancy using systems biology approaches. *Drug Discov. Today* **19:** 1402–1409.
11. Lin LL, Hsia CR, Hsu CL *et al.* (2015) Integrating transcriptomics and proteomics to show that tanshinone IIA suppresses cell growth by blocking glucose metabolism in gastric cancer cells. *BMC Genomics* **16:** 41.
12. Lin LL, Huang HC, and Juan HF. (2012) Discovery of biomarkers for gastric cancer: A proteomics approach. *J. Proteome Res.* **75:** 3081–3097.
13. Chang HY, Huang HC, Huang TC *et al.* (2012) Ectopic ATP synthase blockade suppresses lung adenocarcinoma growth by activating the unfolded protein response. *Cancer Res.* **72:** 4696–4706.

14. Tseng CW, Yang JC, Chen CN *et al.* (2011) Identification of 14-3-3β in human gastric cancer cells and its potency as a diagnostic and prognostic biomarker. *Proteomics* **11:** 2423–2439.

15. Lin LL, Chen CN, Lin WC *et al.* (2008) Annexin A4: A novel molecular marker for gastric cancer with *Helicobacter pylori* infection using proteomics approach. *Proteomics: Clin. Appl.* **2:** 619–634.

16. Huang TC, Huang HC, Chang CC *et al.* (2007) An apoptosis-related gene network induced by novel compound-cRGD in human breast cancer cells. *FEBS Letters* **581:** 3517–3522.

17. Cheng KC, Huang HC, Chen JH *et al.* (2007) *Ganoderma lucidum* polysaccharides in human monocytic leukemia cells: From gene expression to network construction. *BMC Genomics* **8:** 411.

18. Hsu JW, Huang HC, Chen ST *et al.* (2011) *Ganoderma lucidum* polysaccharides induce macrophage-like differentiation in human leukemia THP-1 cells via caspase and p53 activation. *Evid. Based Complement. Alternat. Med.* **2011:** 358717.

19. Lin LL, Huang HC, and Juan HF. (2012) Revealing the molecular mechanism of gastric cancer marker Annexin A4 in cancer cell proliferation using exon arrays. *PLoS ONE* **7:** e44615.

20. Chang HY, Huang TC, Chen NN *et al.* (2014) Combination therapy targeting ectopic ATP synthase and 26s proteasome induces ER stress in breast cancer cells. *Cell Death Dis.* **5:** e1540.

21. Wu YH, Hu CW, Chien CW *et al.* (2013) Quantitative proteomic analysis of human lung tumor xenografts treated with the ectopic ATP synthase inhibitor citreoviridin. *PLoS ONE* **8:** e70642.

22. Juan HF, Wang IH, Huang TC *et al.* (2006) Proteomics analysis of a novel compound — cyclic RGD in breast carcinoma cell line MCF-7. *Proteomics* **6:** 2991–3000.

23. Bartel DP. (2009) MicroRNAs: Target recognition and regulatory functions. *Cell* **136:** 215–233.

24. Hu CW, Tseng CW, Chien CW *et al.* (2013) Quantitative proteomics reveals diverse roles of miR-148a from gastric cancer progression to neurological development. *J Proteome Res.* **12:** 3993–4004.

25. Lin R, Wang WR, Liu JT *et al.* (2006) Protective effect of tanshinone IIA on human umbilical vein endothelial cell injured by hydrogen peroxide and its mechanism. *J. Ethnopharmacol.* **108:** 217–222.

2. Transcriptome Analysis: Library Construction

Hsin-Yi Chang and Hsueh-Fen Juan*

*Institute of Molecular and Cellular Biology,
National Taiwan University, Taipei, Taiwan*
*yukijuan@ntu.edu.tw

1. Introduction

Transcriptome analysis reveals pivotal gene expression regulations in given biological statuses. Conventional microarray analyses provide opportunities to study relevant levels of transcripts with known sequences.[1] Therefore, gene expressions, alternative splicing of genes, and amount of any user-defined sequences can be measured in high-throughput manner. However, several limitations arise while employing microarray analysis: Loss of linearity of signal, limited detection of weak fluorescence intensity, and inability to discover novel transcripts.[2–4]

Next-generation sequencing (NGS) provides digital counts in transcripts, sequence information in single nucleotide resolution, and high sensitive detection in single read. These advantages give rise to a step forward towards revealing new phenotypes of transcriptome. Transcriptome analysis has been successfully applied rapidly in various biological issues, such as gene expression, differential splicing, and allele-specific expression of transcripts. Furthermore, the application of NGS in detecting specific diseases, specific genes or mutations (targeted sequencing) for pathological determination makes the implementation in personalized medicine progressive.[5,6] Moreover, the advantage of NGS in sequencing non-model organisms is to provide the analytical power to reveal gene structures rapidly and molecular ecology discipline.[7,8]

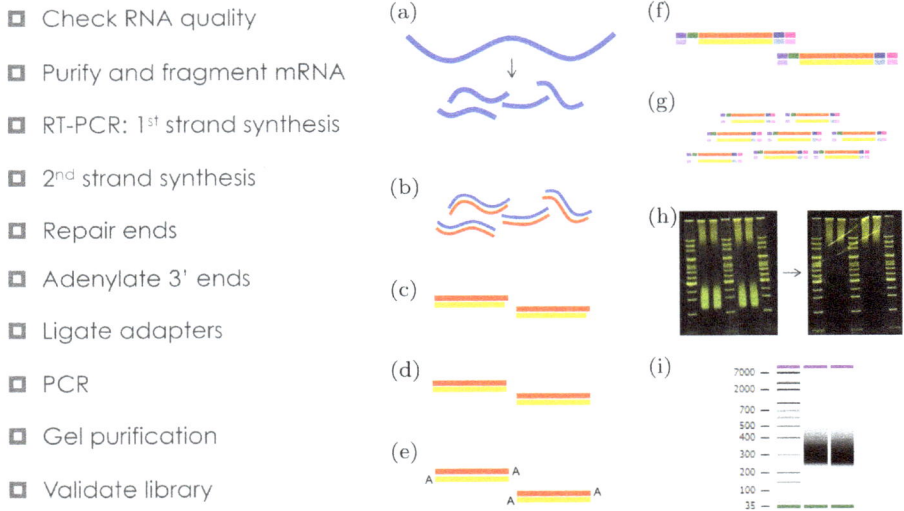

Figure 1. Schematic representation of library construction for transcriptome analysis. The step-by-step protocol is included in this chapter: (a) Purify and fragment RNA, (b) RT-PCR: first strand cDNA synthesis, (c) RT-PCR: second strand cDNA synthesis, (d) repair ends, (e) adenylate 3′ ends, (f) ligate adapters, (g) PCR amplification, (h) Gel purification of amplified library, (i) validate the library.

In the past decade, several parallel NGS platforms have been developed to provide low-cost, high-throughput sequencing.[3,4,9] In this chapter, we describe the sample preparation in detail for the most commonly used platform in research and clinical laboratories, the Illumina platform, as illustrated in Fig. 1.

Consumables and equipment

TruSeq RNA sample Preparation v2 Kit (Illumina, RS-122-2001)

1.5 mL RNase/DNase-free non-sticky tubes (Life Technologies, AM12450)

$10\,\mu L$, $200\,\mu L$, and $1000\,\mu L$ barrier pipette tips (General lab supplier)

$10\,\mu L$, $200\,\mu L$, and $1000\,\mu L$ single channel pipettes (General lab supplier)

Nuclease-free ultra pure water (General lab supplier)

Ethanol 200 proof (absolute) for molecular biology (Sigma-Aldrich, E7023)

RNase/DNase-free PCR tubes (General lab supplier)

RNase/DNase zapper (General lab supplier)

SuperScript II Reverse Transcriptase (Invitrogen, 18064-014)

Agencourt AMPure XP 60 mL kit (Bechman Coulter Genomics, A63881)

Tris-HCl 10 mM, pH8.5 with 0.1% Tween 20 (General lab supplier)

Tween 20 (Sigma-Aldrich, P7949)

Agilent 2100 Bioanalyzer (Agilent Technologies Inc., G2939AA)

Magnetic stand (Life Technologies, 12321D)

Thermal cycler with heated lid (General lab supplier)

Centrifuge (General lab supplier)

Vortexer (General lab supplier)

Preparation

Thaw the following items to room temperature (25°C) for 30 min before use.

— RNA Purification Beads (RPB)	— End Repair Control (CTE)
— Bead Binding Buffer (BBB)	— End Repair Mix (ERP)
— Bead Washing Buffer (BWB)	— A-Tailing Control (CTA)
— Elution Buffer (ELB)	— A-Tailing Mix (ATL)
— Elute, Prime, Fragment Mix (EPF)	— Ligation Control (CTL)
— First Strand Master Mix (FSM)	— RNA Adapters Indices
— Resuspension Buffer (RSB)	— Stop Ligation Buffer (STL)
— Second Strand Master Mix (SSM)	— PCR Master Mix (PMM, on ice)
— AMPure XP Beads	— PCR Primer Cocktail (PPC, on ice)

I. *Check RNA quality*

Before proceeding with library construction, sample RNA quality and integrity should be checked by an Agilent 2100 Bioanalyzer. An RNA integrity number (RIN) of at least 8.0 is the recommended threshold. Figure 2 shows examples of high quality total RNA extracted from human lung adenocarcinoma cell A549 using conventional Trizol protocol and treated with DNase I. Use of 0.1–4 μg of total RNA for transcriptome library construction is recommended.

II. *Purify and fragment mRNA*

- Dilute the total RNA with nuclease-free ultra pure water (NFW) to a final volume of 50 μL in a PCR tube.

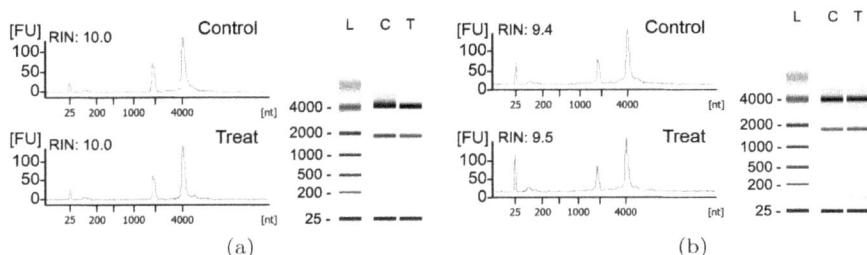

Figure 2.   RNA electropherograms and the digital gel (right) obtained with the Agilent 2100 Bioanalyzer (L, ladder; C, control sample; T, treated sample; 100 ng per sample for all RNAs). Total RNA was (a) extracted using Trizol and (b) subsequently treated with DNase I to move genomic DNA contamination.

- Vortex the RPB tubes vigorously to completely resuspend the oligo-dT beads.
- Add 50 μL of RPB to each sample to bind the poly-A RNA to the oligo-dT magnetic beads. Gently pipette the entire volume up and down 10 times to mix thoroughly.
- Place the sample on the thermal cycler. Close the lid and denature the RNA using the following program.

65° C for 5 minutes, 4° C hold

- Remove the sample from the thermal cycler, centrifuge briefly, and allow to stand on the bench for 5 minutes at room temperature to allow the RNA to bind to the beads.
- Place the sample on the magnetic stand at room temperature for 5 minutes to separate the poly-A RNA bound beads from the solution.
- Remove and discard the supernatant from each sample.
- Remove the sample tube from the magnetic stand and wash the beads with 200 μl of BWB to remove unbound RNA. Gently pipette the entire volume up and down 10 times to mix thoroughly.
- Centrifuge briefly and place the sample on the magnetic stand at room temperature for 5 minutes.
- Remove and discard all of the supernatant from each sample.

Note: The supernatant contains the majority of the ribosomal RNA (rRNA) and other non-mRNA.

- Remove the sample tube from the magnetic stand and add 50 µL of ELB. Gently pipette the entire volume up and down 10 times to mix thoroughly.
- Place the sample on the thermal cycler to elute mRNA from the beads using the following program:

  80°C for 2 minutes, 25°C hold

- Remove the sample tube from the thermal cycler and add 50 µL of BBB. Gently pipette the entire volume up and down 10 times to mix thoroughly.

  Note: This allows mRNA to specifically rebind the beads, while reducing the amount of rRNA that non-specifically binds.

- Incubate the sample at room temperature for 5 minutes.
- Centrifuge briefly and place the sample on magnetic stand at room temperature for 5 minutes.
- Remove and discard all the supernatant from each sample.
- Remove the sample tube from the magnetic stand and wash the beads with 200 µL of BWB. Gently pipette the entire volume up and down 10 times to mix thoroughly.
- Centrifuge briefly and place the sample on magnetic stand at room temperature for 5 minutes.
- Remove and discard all of the supernatant from each sample.

  Note: The supernatant contains residual rRNA and other contaminants that were released in the first elution and did not rebind the beads.

- Remove the sample tube from the magnetic stand and add 19.5 µL of EPF to each sample. Gently pipette the entire volume up and down 10 times to mix thoroughly.

  Note: The EPF contains random hexamers for RT priming and serves as the 1st strand cDNA synthesis reaction buffer.

- Place the sample on the thermal cycler to elute, fragment, and prime the RNA from the beads using the following program:

  94°C for 8 minutes, 4°C hold

- Remove the sample from the thermal cycler, centrifuge briefly, and proceed immediately to 1st strand cDNA synthesis.

III. *RT-PCR: 1st strand synthesis*

- Centrifuge briefly and place the sample on magnetic stand at room temperature for 5 minutes.
- Transfer 17 µL of the supernatant (fragmented and primed mRNA) from each sample to the corresponding new PCR tube.

- Prepare FSM and SuperScript II mix at the ratio of $1\,\mu$L SuperScript II for each $9\,\mu$L FSM. Gently mix thoroughly and centrifuge briefly.

  Note: Return the FSM and SuperScript II to -15 to -25°C storage immediately after use.

- Add $8\,\mu$L of FSM and SuperScript II mix to each sample. Gently pipette the entire volume up and down 10 times to mix thoroughly.
- Place the sample on the thermal cycler to synthesize the 1st strand cDNA using the following program:

$$25^\circ C \ for \ 10 \ minutes$$

$$42^\circ C \ for \ 50 \ minutes$$

$$70^\circ C \ for \ 15 \ minutes$$

$$Hold \ at \ 4^\circ C$$

- Remove the sample from the thermal cycler, centrifuge briefly, and proceed immediately to 2nd strand cDNA synthesis.

IV. *RT-PCR: 2nd strand synthesis*

- Pre-heat the thermal cycler to 16°C.
- Add $25\,\mu$L of SSM to each sample. Gently pipette the entire volume up and down 10 times to mix thoroughly.
- Place the sample on the thermal cycler to synthesize the 2nd strand cDNA.

$$16^\circ C \ for \ 60 \ minutes$$

- Vortex the AMPure XP beads until they are well dispersed, then add $90\,\mu$L of well-mixed AMPure XP beads to each sample containing $50\,\mu$L of double stranded cDNA. Gently pipette the entire volume up and down 10 times to mix thoroughly.
- Incubate the sample at room temperature for 15 minutes.
- Centrifuge briefly and place the sample on magnetic stand at room temperature for 5 minutes to make sure that all beads are bound to the side of the wells.
- Remove and discard $135\,\mu$L of supernatant from each sample.

  Note: Leave the sample tubes on the magnetic stand while performing all the following 80% EtOH wash steps.

- With the sample remaining on the magnetic stand, add $200\,\mu$L of freshly prepared 80% EtOH to each sample without disturbing the beads.

- Incubate the sample at room temperature for 30 seconds, then remove and discard all of the supernatant from each well.
- Repeat the 80% EtOH wash steps twice.
- Open the lid and let the sample tube stand at room temperature for 15 minutes to dry and then remove the sample from the magnetic stand.
- Add $52.5\,\mu$L RSB to each sample. Gently pipette the entire volume up and down 10 times to mix thoroughly.
- Incubate the sample at room temperature for 2 minutes.
- Centrifuge briefly and place the sample on the magnetic stand at room temperature for 5 minutes.
- Transfer $50\,\mu$L of the supernatant (ds cDNA) to a new PCR tube.

  Note: The purified ds DNA can be stored at -15 to $-25°$C for up to seven days.

V. *Repair ends*

- Pre-heat the thermal cycler to $30°$C.
- Prepare the in-line control reagent by diluting the CTE to $1/100$ in RSB ($1\,\mu$L CTE $+ 99\,\mu$L RSB) before use. Add $10\,\mu$L of dilute CTE to each sample that contains $50\,\mu$L of dsDNA. If not using the in-line control reagent, add $10\,\mu$L of RSB instead.
- Add $40\,\mu$L of ERP to each sample. Gently pipette the entire volume up and down 10 times to mix thoroughly.
- Place the sample on the thermal cycler to pre-form end repair.

30°C for 30 minutes

- Vortex the AMPure XP beads until they are well dispersed, then add $160\,\mu$L of well-mixed AMPure XP beads to each sample containing $100\,\mu$L of end repaired dsDNA mixture. Gently pipette the entire volume up and down 10 times to mix thoroughly.
- Incubate the sample at room temperature for 15 minutes.
- Centrifuge briefly and place the sample on magnetic stand at room temperature for 5 minutes to make sure that all beads are bound to the side of the wells.
- Remove and discard $127.5\,\mu$L of supernatant from each sample.

  Note: Leave the sample tubes on the magnetic stand while performing all the following 80% EtOH wash steps.

- With the sample remaining on the magnetic stand, add $200\,\mu$L of freshly prepared 80% EtOH to each sample without disturbing the beads.

- Incubate the sample at room temperature for 30 seconds, then remove and discard all of the supernatant from each well.
- Repeat the 80% EtOH wash steps twice.
- Open the lid and let the sample tube stand at room temperature for 15 minutes to dry and then remove the sample from the magnetic stand.
- Add $17.5\,\mu$L RSB to each sample. Gently pipette the entire volume up and down 10 times to mix thoroughly.
- Incubate the sample at room temperature for 2 minutes.
- Centrifuge briefly and place the sample on the magnetic stand at room temperature for 5 minutes.
- Transfer $15\,\mu$L of the clear supernatant to a new PCR tube.

  Note: The purified ds DNA can be stored at $-15 - -25°$C for up to 7 days.

VI. *Adenylate 3′ ends*

- Pre-heat the thermal cycler to 37°C.
- Prepare the in-line control reagent by diluting the CTA to 1/100 in RSB ($1\,\mu$L CTA + $99\,\mu$L RSB) before use. Add $2.5\,\mu$L of dilute CTA to each sample that contains $15\,\mu$L of ds DNA. If not using the in-line control reagent, add $2.5\,\mu$L of RSB instead.
- Add $12.5\,\mu$L of ATL to each sample. Gently pipette the entire volume up and down 10 times to mix thoroughly.
- Place the sample on the thermal cycler to pre-form end repair.

  37° C for 30 minutes

- Remove the sample from the thermal cycler, centrifuge briefly, and proceed immediately to Adapter Ligation.

VII. *Ligate adapters*

- Pre-heat the thermal cycler to 30°C.
- Prepare the in-line control reagent by diluting the CTL to 1/100 in RSB ($1\,\mu$L CTL + $99\,\mu$L RSB) before use. Add $2.5\,\mu$L of dilute CTL to each sample. If not using the in-line control reagent, add $2.5\,\mu$L of RSB instead.
- Before use, immediately remove the Ligation Mix tube from $-15°$C to $-25°$C storage. Add $2.5\,\mu$L of Ligation Mix to each sample. Gently pipette the entire volume up and down 10 times to mix thoroughly.
- Return the Ligation Mix tube back to $-15°$C to $-25°$C storage immediately after use.
- Add $2.5\,\mu$L of appropriate/desired thawed RNA Adapter Index to each sample. Gently pipette the entire volume up and down 10 times to mix thoroughly.

- Place the sample on the thermal cycler to preform adapter ligation.

30° C for 10 minutes

- Remove samples from thermal cycler. Add 5 µL of STL to each sample to inactivate the ligation. Gently pipette the entire volume up and down 10 times to mix thoroughly.
- Vortex the AMPure XP beads until they are well dispersed, then add 42 µL of well-mixed AMPure XP beads to each sample containing 42.5 µL of sample. Gently pipette the entire volume up and down 10 times to mix thoroughly.
- Incubate the sample at room temperature for 15 minutes.
- Centrifuge briefly and place the sample on magnetic stand at room temperature for 5 minutes to make sure that all beads are bound to the side of the wells.
- Remove and discard 79.5 µL of supernatant from each sample.

  Note: Leave the sample tubes on the magnetic stand while performing all the following 80% EtOH wash steps.

- With the sample remaining on the magnetic stand, add 200 µL of freshly prepared 80% EtOH to each sample without disturbing the beads.
- Incubate the sample at room temperature for 30 seconds, then remove and discard all of the supernatant from each well.
- Repeat the 80% EtOH wash steps twice.
- Open the lid and let the sample tube stand at room temperature for 15 minutes to dry and then remove the sample from the magnetic stand.
- Add 52.5 µL RSB to each sample. Gently pipette the entire volume up and down 10 times to mix thoroughly.
- Incubate the sample at room temperature for 2 minutes.
- Centrifuge briefly and place the sample on the magnetic stand at room temperature for 5 minutes.
- Transfer 50 µL of the clear supernatant to a new PCR tube.
- Vortex the AMPure XP beads until they are well dispersed, then add 50 µL of well-mixed AMPure XP beads to each sample containing 50 µL of sample. Gently pipette the entire volume up and down 10 times to mix thoroughly.
- Incubate the sample at room temperature for 15 minutes.
- Centrifuge briefly and place the sample on magnetic stand at room temperature for 5 minutes to make sure that all beads are bound to the side of the wells.
- Remove and discard 95 µL of supernatant from each sample.

Note: Leave the sample tubes on the magnetic stand while performing all the following 80% EtOH wash steps.

- With the sample remaining on the magnetic stand, add $200\,\mu$L of freshly prepared 80% EtOH to each sample without disturbing the beads.
- Incubate the sample at room temperature for 30 seconds, then remove and discard all of the supernatant from each well.
- Repeat the 80% EtOH wash steps twice.
- Open the lid and let the sample tube stand at room temperature for 15 minutes to dry and then remove the sample from the magnetic stand.
- Add $22.5\,\mu$L RSB to each sample. Gently pipette the entire volume up and down 10 times to mix thoroughly.
- Incubate the sample at room temperature for 2 minutes.
- Centrifuge briefly and place the sample on the magnetic stand at room temperature for 5 minutes.
- Transfer $20\,\mu$L of the clear supernatant to a new PCR tube.

    Note: The purified DNA can be stored at $-15 - -25^\circ$C for up to 7 days.

- (Optional) Transfer $1\,\mu$L of the clear supernatant to a new PCR tube for library validation.

VIII. *PCR amplification*

- Add $5\,\mu$L of thawed PPC to each sample. Gently pipette the entire volume up and down 10 times to mix thoroughly.
- Add $25\,\mu$L of thawed PMM to each sample. Gently pipette the entire volume up and down 10 times to mix thoroughly.
- Place the sample on the thermal cycler to amplify the adapter ligated DNA fragments using the following program:

$$98^\circ C \text{ for } 30 \text{ seconds}$$
$$13 \text{ Cycles of}:$$
$$-98^\circ C \text{ for } 10 \text{ seconds}$$
$$-60^\circ C \text{ for } 30 \text{ seconds}$$
$$-72^\circ C \text{ for } 30 \text{ seconds}$$
$$72^\circ C \text{ for } 5 \text{ minutes}$$
$$Hold \text{ at } 10^\circ C$$

- Vortex the AMPure XP beads until they are well dispersed, then add $50\,\mu$L of well-mixed AMPure XP beads to each sample containing $50\,\mu$L of the PCR amplified library. Gently pipette the entire volume up and down 10 times to mix thoroughly.

- Incubate the sample at room temperature for 15 minutes.
- Centrifuge briefly and place the sample on magnetic stand at room temperature for 5 minutes to make sure that all beads are bound to the side of the wells.
- Remove and discard 95 μL of supernatant from each sample.
  Note: Leave the sample tubes on the magnetic stand whiling performing all the following 80% EtOH wash steps.

- With the sample remaining on the magnetic stand, add 200 μL of freshly prepared 80% EtOH to each sample without disturbing the beads.
- Incubate the sample at room temperature for 30 seconds, then remove and discard all of the supernatant from each well.
- Repeat the 80% EtOH wash steps twice.
- Open the lid and let the sample tube stand at room temperature for 15 minutes to dry and then remove the sample from the magnetic stand.
- Add 32.5 μL RSB to each sample. Gently pipette the entire volume up and down 10 times to mix thoroughly.
- Incubate the sample at room temperature for 2 minutes.
- Centrifuge briefly and place the sample on the magnetic stand at room temperature for 5 minutes.
- Transfer 30 μL of the clear supernatant to a new PCR tube.

Note: The purified DNA library can be stored at -15 to $-25°$C for up to seven days.

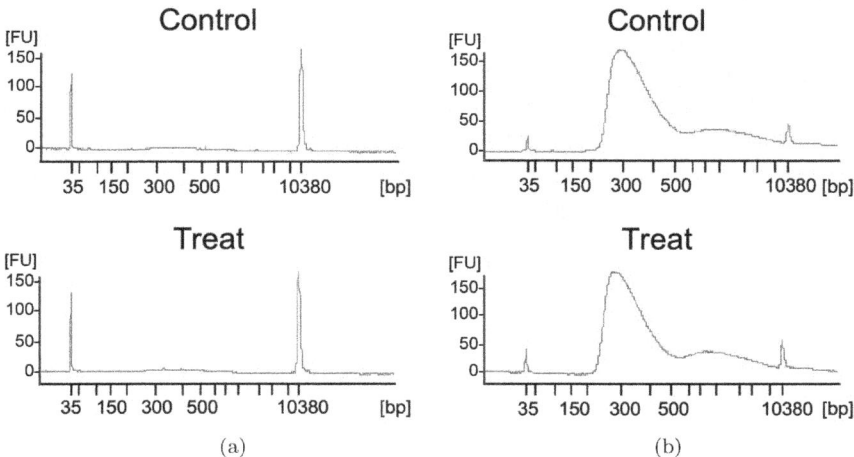

Figure 3. Example of TruSeq RNA Sample Prep v2 Library Size Distribution. Electropherograms were obtained with the Agilent 2100 Bioanalyzer (Control, control sample; Treat, treated sample) before (a) and after (b) library amplification.

- (Optional) Transfer 1 μL of the clear supernatant to a new PCR tube for library validation.

IX. *Validate library*

- Load 1 μL of the clear supernatant from the ligated DNA samples and the PCR amplified DNA samples on an Agilent Technologies 2100 Bioanalyzer using a DNA-specific chip.
- Check the size and purity of the sample. The final product should be a band at approximately 260 bp as shown in Fig. 3.

References

1. Su Z, Li Z, Chen T *et al.* (2011) Comparing next-generation sequencing and microarray technologies in a toxicological study of the effects of aristolochic acid on rat kidneys. *Chem. Res. Toxicol.* **24:** 1486–1493.
2. Koboldt DC, Steinberg KM, Larson DE *et al.* (2013) The next-generation sequencing revolution and its impact on genomics. *Cell* **155:** 27–38.
3. Mardis ER. (2013) Next-generation sequencing platforms. *Annu. Rev. Anal. Chem.* **6:** 287–303.
4. Mutz KO, Heilkenbrinker A, Lönne M *et al.* (2013) Transcriptome analysis using next-generation sequencing. *Curr. Opin. Biotechnol.* **24:** 22–30.
5. Meldrum C, Doyle MA, and Tothill RW. (2011) Next-generation sequencing for cancer diagnostics: A practical perspective. *Clin. Biochem. Rev.* **32:** 177.
6. Rabbani B, Mahdieh N, Hosomichi K *et al.* (2012) Next-generation sequencing: Impact of exome sequencing in characterizing mendelian disorders. *Clin. Biochem. Rev.* **57:** 621–632.
7. Ekblom R and Galindo J. (2011) Applications of next generation sequencing in molecular ecology of non-model organisms. *Heredity* **107:** 1−15.
8. Berkman PJ, Lai K, Lorenc MT *et al.* (2012) Next-generation sequencing applications for wheat crop improvement. *Am. J. Bot.* **99:** 365–371.
9. van Dijk EL, Auger H, Jaszczyszyn Y *et al.* (2014) Ten years of next-generation sequencing technology. *Trends Genet.* **30:** 418–426.

3. Quantitative Proteome: The Isobaric Tags for Relative and Absolute Quantitation (iTRAQ)

Yi-Hsuan Wu and Hsueh-Fen Juan*

*Institute of Molecular and Cellular Biology,
National Taiwan University, Taipei, Taiwan*
**yukijuan@ntu.edu.tw*

1. Introduction to the methods for quantitative proteome

The term "proteome" was first introduced in 1996 in analog to genome with a definition of the entire complements of proteins expressed in a specific state of a cell, tissue or an organism.[1] Despite the similarity among genome, transcriptome and proteome, the proteomic profiling of a subject cannot be directly transferred from its genome or transcriptome. The events of gene regulation, alternative splicing variants, post-transcriptional and post-translational modifications are responsible for the diverse profiling of proteome that are distinct from genome or transcriptome. The field of proteomics not only identifies and quantifies proteins but also encompasses the dimensions of protein subcellular localizations, post-translational modifications, temporal changes in expression and interactions.[2]

The development of mass spectrometry (MS) greatly drives the advancement of proteomic studies. Different from single-stage MS measuring the mass of polypeptides, tandem mass spectrometry (MS/MS) contains the determination of mass as single-stage MS and a fragmentation of selected peptides by collision for further analysis. In most cases, MS/MS is performed to determine additional structure information of the peptides including amino acid sequence.[3] Shotgun proteomics, coined in 1998, refers to the MS-based proteomics that determines proteins by measuring the peptides derived from the digestion of the proteins.[4] Shotgun proteomics, also called peptide-centric or "bottom-up" proteomics, is widely used for proteomic investigation. The aim of proteomics is to accurately characterize as many proteins as possible, and the ultimate goal is the explicit information of the complex mixture containing at least 10,000 different proteins in a cell population.[5]

The generic workflow of shotgun proteomics starts from the digestion of proteins into peptides by sequence-specific enzymes such as trypsin. The peptides are subsequently separated by on-line Ultra Performance Liquid Chromatography (UPLC) and converted into gas phase ions by electrospray.[6] The peptide ions are scanned by mass spectrometer to generate mass spectra, and selected peptides from the mass spectra are subjected to collision-induced dissociation (CID) for fragmentation. The fragment ions are measured by the second mass spectrometer, and the fragment ion spectra generated are assigned to their corresponding peptide by sequence database search algorithm. Finally, protein inference is performed by assigning peptide sequences to proteins.

Quantitative proteomics is the identification of proteins and measurement of protein abundance in different samples. The profiling from quantitative proteomics enables us to compare the proteomes of samples in different biological status, such as altered expression of proteins in response to a specific stimulus. The proteins showing changes in abundance are possibly associated with the biological processes that determine the different status of the samples. Two major methods are available for performing quantitative analysis of MS-based shotgun proteomics, the label-free quantification and stable isotope-based quantification.[7] The label-free quantification can compare across many samples in one experiment, but the accuracy is lower and not sensitive enough to small changes in abundance. The isotope-based quantification has the advantage of accuracy, but it is only used for comparison of up to eight samples at a time. The data of quantitative proteomics have two forms, the absolute amount of each protein in a sample or the relative change in protein amount among different samples. The absolute quantification or AQUA[8] is achieved by comparing the signal of the peptides from the protein of interest to a spiked in or isotope-labeled peptide. The relative quantification is more well established compared to the absolute one that required time-consuming processes of developing reference materials and assay condition for proteins.[9]

One of the label-free quantification is achieved by spectrum counts, inferring protein abundance by the number of times a peptide is observed and the number of distinct peptides observed from a given protein. Another label-free method is quantification by comparing the intensity of mass spectrometric signal of each peptide from a given protein.[7] Isotope-based methods determine protein abundance by incorporating specific molecules into peptides by metabolic or chemical labeling. Samples are labeled with different versions of a chemical reagent and can be combined, resulting in paired peptides with a defined mass difference. The ratio of signal intensities from the paired peptides with different masses can be determined after the combined samples are analyzed.[9] Stable isotope labeling by amino acids in cell culture (SILAC)[10] is one of the powerful approaches by metabolic labeling. The cells are cultured in media containing stable ^{13}C or ^{15}N isotope-labeled arginine and lysine, so the labeled amino acids are incorporated into each protein in the cells. In chemical labeling, the isotope-coded affinity tag (ICAT)[11] labeled the cysteine residue in peptides with a reagent containing eight deuteriums and a biotin group for purification of labeled peptides. Several other strategies of chemical labeling targeting on amino acid are also available: methylation,[12] esterification, the isotope-coded protein label (ICPL),[13] and the tandem mass tags (TMT).[14]

The isobaric tags for relative and absolute quantitation (iTRAQ)[15] is one of the most popular chemical labeling methods. The iTRAQ reagent is amide reactive and can link to the N-terminus and lysine side chains of peptides (Fig. 1). The term "isobaric" describes the characteristic that different iTRAQ reagents labeling different samples have equal mass. The mass is not distinguished in the survey scan of the first MS. The iTRAQ reagents are fragmented by CID before the second MS and reporter ions of m/z 114, 115, 116 and 117 in MS/MS spectra for four-plexed reagent set are generated. In the MS/MS spectra, the intensities of the reporter ions are used for peptide quantification while the remaining peaks are used for peptide sequence identification. One of the strengths of iTRAQ strategy is the enhanced signals of peptides by combining all labeled-samples into a single analysis, greatly increasing the accuracy of quantification.[16]

The iTRAQ quantitative proteomics is a well-established method suitable for a multiplex experiment. The procedures of preparing samples for iTRAQ quantitative proteomic analysis include protein extraction, reduction, alkylation, digestion, iTRAQ-labeling, strong cation exchange (SCX) chromatography and ZipTip desalting (Fig. 2). The detailed protocol of sample preparation for LC-MS/MS analysis is described later in this chapter. The methods of gel-assisted digestion and gel extraction[17] are applied in this protocol for the protein digestion step. Please also refer to the reference guide of the iTRAQ Reagents to note the substances that may interfere with the

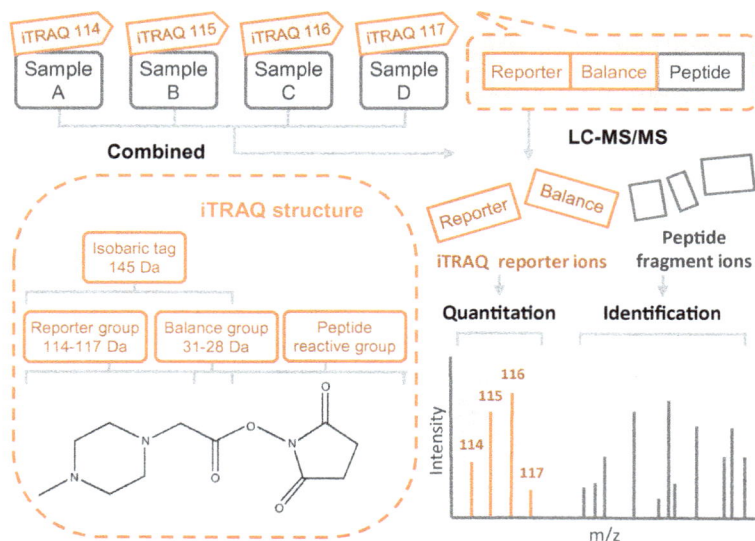

Figure 1.   Structure and workflow of iTRAQ.

protocol of iTRAQ method prior to beginning the experiment. The possible interference of those substances is reaction with iTRAQ reagent, inactivation of trypsin digestion and interference with protein reduction.

2.   Protocol of sample preparation for iTRAQ quantitative proteomics

Materials

Reagents

- Tris-base
- 6M HCl
- 10% Sodium dodecyl sulfate (SDS)
- Glycerol
- Protease inhibitor
- Liquid nitrogen
- Pierce BCA Protein Assay Kit (Pierce cat. no. 23225)
- Triethylammonium bicarbonate buffer (TEAB) (Sigma cat. no. T7408)
- Tris(2-carboxyethyl)phosphine hydrochloride (TCEP) (Sigma cat. no. C4706)
- *S*-Methyl methanethiosulfonate (MMTS) (Sigma cat. no. 208795)

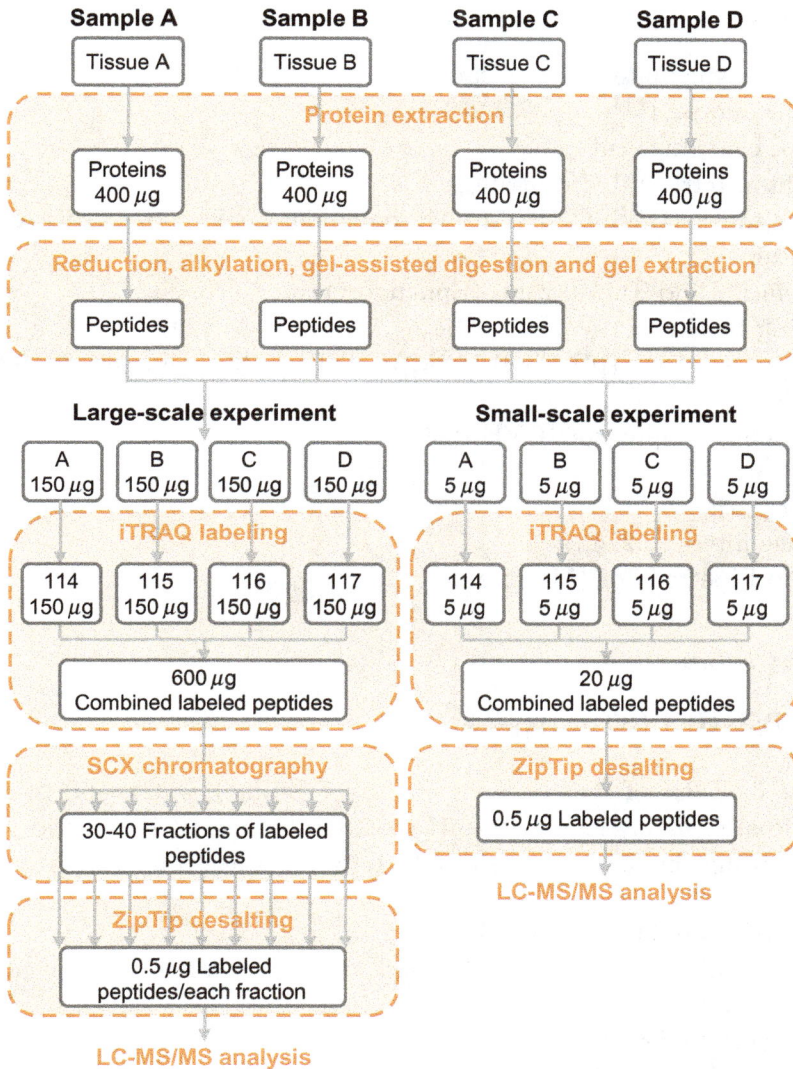

Figure 2. Flowchart of sample preparation for iTRAQ quantitative proteomics. The detailed description and protocols are shown below.

- Acrylamide/bis-acrylamide (40%, v/v, 37.5:1)
- Ammonium persulfate (APS)
- N,N,N',N'-Tetramethylethylenediamine (TEMED)
- Acetonitrile (ACN)
- Sequencing grade modified trypsin (Promega, cat. no. V5111)
- Trifluoroacetic acid (TFA)
- iTRAQ Reagents Multiplex Kit (SCIEX, cat. no. 4352135)

Equipment

- Magnetic stir plate and stir bar
- Mortar and pestle
- 1.5 mL centrifuge tubes
- Benchtop refrigerated centrifuge
- Homogenizer (LABSONIC M ultrasonic homogenizer; Sartorius AG)
- Vortexer
- Protein LoBind Tubes, 2 mL (Eppendorf)
- Heating block
- pH-Fix test strips (pH-Fix 0-14, MACHEREY-NAGEL, cat. no. 92110)
- Tweezers
- Centrifugal evaporator (CVE-2000; Eyela)
- Parafilm
- Water bath
- Syringe filters (0.2 μm)
- ZipTip Pipette Tips (Millipore)

Reagent setup

- **1 M Tris-HCl (30 mL)**

Dissolve 3.6342 g of Tris-base in 10 mL of Milli-Q water and stir on a magnetic stir plate. Adjust the pH to 6.8 with 6M HCl and add Milli-Q water to bring the final volume to 30 mL. Store it at room temperature.

- **100 mM Tris-HCl (30 mL)**
  Mix 3 mL of 1 M Tris-HCl with 27 mL of Milli-Q water. Store it at room temperature.
- **1 × Lysis buffer (20 mL)**

10% Sodium dodecyl sulfate (SDS)	2 mL
100 mM Tris HCl	10 mL
Glycerol	2 mL
Milli-Q water	6 mL

Aliquot 1 mL and store at −30°C. Add protease inhibitor just before use.
- **200 mM Triethylammonium bicarbonate buffer (TCEP) (200 μL)**
  Dissolve 11.45 mg of TCEP in 200 μL of Milli-Q water. Freshly prepare this reagent and keep it at 4°C before the experiment.

- **200 mM S-Methyl methanethiosulfonate (MMTS) (500 μL)**
  Mix 9.73 μL of MMTS and 490.27 μL of Milli-Q water. Store at 4°C and avoid exposure to light.
- **10% Ammonium persulfate (APS) (1 mL)**
  Dissolve 0.1 g of APS powder in 1 mL of Milli-Q water. Store at −20°C.
- **25 mM Triethylammonium bicarbonate buffer (TEAB) (50 mL)**
  Mix 1.25 mL of 1 M TEABC and 48.75 mL of Milli-Q water. Store it at room temperature.
- **25 mM TEAB/50% Acetonitrile (ACN) (50 mL)**
  Mix 1.25 mL of 1 M TEABC, 25 mL of 100% ACN and 23.75 mL of Milli-Q water. Store it at room temperature.
- **10% Trifluoroacetic acid (TFA) (10 mL)**
  Mix 1 mL of 100% TFA and 9 mL of Milli-Q water. Store at room temperature.
- **0.1% TFA (50 mL)**
  Mix 0.5 mL of 10% TFA and 49.5 mL of Milli-Q water. Store it at room temperature.
- **50% ACN/0.1% TFA (50 mL)**
  Mix 25 mL of 100% ACN, 0.5 mL of 10% TFA and 24.5 mL of Milli-Q water. Store it at room temperature.

Procedure

Protein extraction from tissue samples

1. Grind the tissue into powder with mortar and pestle in liquid nitrogen. Add liquid nitrogen frequently to ensure that the tissue does not thaw during grinding. Distribute the powder to 1.5 mL centrifuge tubes. For long-term storage, this tissue sample can be stored at −80°C.
2. Precool the centrifuge and rotor to 4°C before starting lysis. Prepare the lysis buffer with volume of about 3–5 times of the volume of tissue sample.

   - Be sure to add protease inhibitor before using the lysis buffer.

3. Add lysis buffer to tissue sample and pipetting or vortex to mix thoroughly.
4. Homogenize the sample solution on ice using homogenizer with 60% amplitude, cycle of 0.6 (operated 0.6 second every second) for 4–5 minutes several times until the solution become almost transparent and not viscous.

5. Centrifuge the lysate at 17,000 g, 4°C for 30 minutes. Transfer the supernatant, which contain the crude extracted proteins, to a new 1.5 mL centrifuge tube. The protein sample can be stored at −30°C for 2–3 days or stored at −80°C for months.

Reduction and alkylation of proteins[17]

1. Freshly prepare 200 mM TCEP and keep it at 4°C before starting the following steps.
2. Measure the concentration of protein with Pierce BCA Protein Assay Kit.
3. Transfer the amount of proteins to a new 2 mL protein lobind tube. The amount of 10−100 μg/sample is suitable for testing or small-scale experiment, while 400 μg/sample is enough for large-scale experiment.
4. Adjust each sample to have equivalent amount and volume with lysis buffer.
5. Add 1 M TEAB to a final concentration of 50 mM TEAB in each sample.
6. (Optional) Examine the pH with pH test strip. The pH of each sample should be about pH 8.5.
7. Protein reduction by adding 200 mM TCEP to a final concentration of 5 mM TCEP in each sample. Incubate the sample at 37°C on a heat block for 30 minutes.
8. Protein alkylation b adding 200 mM MMTS to a final concentration of 2 mM MMTS in each sample. Incubate the sample at room temperature for 30 minutes in the dark.

Gel-assisted digestion of proteins[17]

1. Add appropriate amount of 40% acrylamide/bis-acrylamide (37.5:1), 10% APS and TEMED sequentially. The volume of each reagent should be in the following proportion:

Reagent	Proportion in volume
Sample	14
40% Acrylamide/bis-acrylamide (37.5:1)	5
10% APS	0.3
TEMED	0.3

- The step of adding TEMED must be done rapidly. Vortex and spin down immediately after adding TEMED to sample. Let it stand at room temperature for 10 minutes to allow gel polymerization.

2. Cut the gel into small pieces with tweezers (about 1 mm in diameter) and add 200 µL of 25 mM TEAB to each sample. Do not remove the tweezers from the tube until the next step.
3. Wash the gel pieces following the below steps:

   (1) Add 25 mM TEAB/50% ACN to a final of 1 mL in each sample. Make sure that no gel pieces remain on the tweezers and remove the tweezers from the tube.

   (2) Vortex the sample for 10 minutes and spin down. Discard the supernatant and let the level of remaining supernatant be the same as the level of gel pieces.

   (3) Repeat step 1 to step 2 several times until no bubble forms immediately after vortexing. If the gel pieces shrink greatly, then proceed to the next step.

   (4) Add 25 mM TEAB to a final of 1 mL in each sample.

   (5) Vortex the sample for 10 minutes and centrifuge at 13,000 for 30 seconds. Discard the supernatant and let the level of remaining supernatant be the same as the level of gel pieces. The gel pieces are harder to precipitate in the 25 mM TEAB than in 25 mM TEAB/50% ACN.

   (6) Repeat steps 1 to 5 until no bubble forms immediately after vortexing in step 5. It is easier to form bubble when washing with 25 mM TEABC than with 25 mM TEABC/50% ACN.

   (7) Dehydrate the gel pieces by adding 100% ACN to a final of 1 mL in each sample. Vortex erectly until the gel pieces become white color and spin down. Discard the supernatant and let the level of remaining supernatant be the same as the level of gel pieces.

   (8) Repeat step 7 until the gel pieces aggregate.

   - Example of the gel-washing procedures:

Reagent	Times
25 mM TEABC/50% ACN	2
25 mM TEABC	1
25 mM TEABC/50% ACN	2
25 mM TEABC	1
25 mM TEABC/50% ACN	1
25 mM TEABC	2
100% ACN	2–3

4. Discard all liquid in the tube and leave the aggregated gel pieces only.
5. Dry the gel pieces with centrifugal evaporator for about 20 minutes.

6. Prepare the trypsin solution:
   (1) If you will exhaust the whole bottle containing $20\,\mu g$ trypsin, then add $200\,\mu L$ of $25\,mM$ TEAB to dissolve trypsin powder.
   (2) If you will use a part of the bottle of trypsin, then add $25\,\mu L$ of the trypsin resuspension dilution buffer supplied in the kit to dissolve trypsin powder.

   - The reagent for dissolving trypsin must be precooled at $4°C$.

7. Add appropriate amount of trypin to the dehydrated gel pieces and allow to stand on ice for a while to let the gel pieces absorb the trypsin solution. The amount of protein chosen initially to the amount of trypsin is 10:1 (g/g).
8. Add $25\,mM$ TEAB to rehydrate the gel pieces. The volume of $25\,mM$ TEAB is 2.5 times the volume of the gel pieces, which is determined by the initial volume of sample and the volume of acrylamide/bis-acrylamide, APS and TEMED added previously.
9. Let the sample stand on ice for 10 minutes and observe the level of $25\,mM$ TEAB in the tube. The level of $25\,mM$ TEAB must be about 5 mm higher than the level of gel pieces.
10. If the amount of $25\,mM$ TEAB is not enough, then add additional amount of $25\,mM$ TEAB to let the reagent cover the gel pieces. Let the sample stand for another 10 minutes and observe the level of $25\,mM$ TEAB again.
11. Repeat step 9 to make sure the $25\,mM$ TEAB is enough to cover the gel pieces.
12. Cap and seal the $2\,mL$ protein lobind tube containing sample with parafilm carefully.
13. Incubate the tube in a water bath at $37°C$ overnight (16–20 hours).

Gel extraction to obtain peptides[17]

1. Prepare several new $2\,mL$ protein lobind tube as final peptide tubes. Prepare several new $1.5\,mL$ centrifuge tubes, some as mixture (of gel pieces and peptides) collection tubes and some as peptide collection tubes.
2. Spin down the sample from the water bath. Transfer the supernatant to the mixture collection tube and let the level of remaining supernatant be the same as the level of gel pieces.
3. Add $200\,\mu L$ of $25\,mM$ TEAB to the gel pieces.

4. Vortex the gel pieces for 15 minutes and spin down. Transfer the supernatant to the mixture collection tube.

   - The volume of 25 mM TEAB added to the gel pieces varies with the volume of the gel pieces.

5. Centrifuge the mixture collection tube at 13,000 g for 30 seconds. Transfer the supernatant to the peptide collection tube.

   - The mixture collection tube may contain small gel pieces. Transfer the supernatant carefully to ensure no gel piece is simultaneously transferred to the peptide collection tube. It is fine to leave a small amount of supernatant in the mixture collection tube.

6. Add 400 μL 0.1% TFA to the gel pieces. Repeat steps 4 to 5.
7. Add 600 μL 50% ACN/0.1% TFA to the gel pieces. Repeat steps 4 to 5.
8. Add 600 μL 100% ACN to the gel pieces. Vortex erectly for 15 minutes until the gel pieces become white color and aggregated. Transfer the supernatant to the mixture collection tube.
9. Add appropriate amount of 100% ACN to the mixture collection tube. Vortex erectly for 15 minutes until the gel pieces become white color and aggregated. Transfer the supernatant to the peptide collection tube.
10. Filter the solution in the peptide collection tube with a syringe filter of 0.2 μm. Collect the filtered solution in the final peptide tube.

    - We recommend to collect only 1 mL solution in each final peptide tube to avoid the solution from leaking out of the tube. Based on the volume of solution used for gel extraction, there may be multiple final peptide tubes.

11. Dry the extracted peptides in final peptide tubes with centrifugal evaporator.

    - Be sure that no ACN is left in the tubes. Do not make the extracted peptides too dry, or it will not be easy to redissolve the peptides.
    - Do not discard the gel pieces until finishing the whole experiment.

iTRAQ tag labeling of peptides

1. Calculate the volume of iTRAQ Dissolution Buffer (supplied in the iTRAQ Reagents Multiplex Kit) needed to be added into the dried peptide to make it a final concentration of about 1−1.5 μg/μL based on the initial amount of protein chosen.

2. Resuspend the dried peptides in each final peptide tube with iTRAQ Dissolution Buffer. Combine the peptide solution in each final peptide tube and the volume of the combined solution should be as calculated in step 1.
3. Vortex the sample for several minutes to fully dissolve the peptides in the buffer.
4. Examine the pH with pH test strip. The pH of each sample should be about pH 8.5.
5. The peptide solution can be stored at $-30°C$ for a few days.

Small-scale experiment: test condition and reproducibility

- The amount of peptides required for iTRAQ labeling is $5\,\mu g$ of peptides for each sample.

1. Measure the concentration of peptides with Pierce BCA Protein Assay Kit.
2. Transfer the appropriate amount of peptides to a new $2\,mL$ protein lobind tube.
3. The amount of peptides in different samples should be equal for comparison. Adjust each sample to have equivalent amount and volume with iTRAQ Dissolution Buffer.
4. Label the tube with the name of the sample and iTRAQ isobaric tag number intended to label the sample.
5. Bring the iTRAQ Reagent vials supplied in the iTRAQ Reagents Multiplex Kit to room temperature.
6. Add $70\,\mu L$ absolute ethanol supplied in the iTRAQ Reagents Multiplex Kit to each iTRAQ Reagent vial. Vortex the mixture for 1 minute and spin down.
7. Calculate the amount of iTRAQ Reagent needed to label each sample. Each iTRAQ Reagent vial is able to label up to $75\,\mu g$ of peptides. Be sure that the percentage of ethanol in each sample mixed with iTRAQ Reagent must be over 50%.
8. Equal amount of peptides from different samples are labeled by adding the calculated amout of iTRAQ Reagent 114, iTRAQ Reagent 115, iTRAQ Reagent 116, or iTRAQ Reagent 117.
9. Cap and seal the $2\,mL$ protein lobind tube containing sample with parafilm carefully.
10. Vortex erectly at room temperature for 1 hour.

- Do not vortex too vigorously.

11. Prepare a new 2 mL protein lobind tube. Spin down the labeled peptides and combine the four samples of labeled peptides to the new tube.

    - Do not combine the labeled peptides to one of the sample tube to ensure the amount of peptides of each sample in the combined tube is equal.

12. Dry the labeled peptides with centrifugal evaporator.

    - No need to make the sample too dry. Proceed to the next step when the labeled peptides become as a small drop of brown liquid.

Large-scale experiment: protein quantification

- The amount of peptides required for iTRAQ labeling is 150 μg of peptides for each sample.

1. Measure the concentration of peptides with Pierce BCA Protein Assay Kit.

2. Transfer the appropriate amount of peptides to a new 2 mL protein lobind tube.

3. The amount of peptides in different samples should be equal for comparison. Adjust each sample to have equivalent amount and volume with iTRAQ Dissolution Buffer.

4. Label the tube with the name of the sample and iTRAQ isobaric tag number intended to label the sample.

5. Bring the iTRAQ Reagent vials supplied in the iTRAQ Reagents Multiplex Kit to room temperature.

6. Add 70 μL absolute ethanol supplied in the iTRAQ Reagents Multiplex Kit to each iTRAQ Reagent vial. Vortex the mixture for 1 minute and spin down.

7. Calculate the amount of iTRAQ Reagent needed to label each sample. Each iTRAQ Reagent vial is able to label up to 75 μg of peptides. Be sure that the percentage of ethanol in each sample mixed with iTRAQ Reagent must be over 50%.

    - Two vials of iTRAQ Reagent are required to label 150 μg of peptides in each sample.

8. Equal amount of peptides from different samples are labeled by adding the calculated amount of iTRAQ Reagent 114, iTRAQ Reagent 115, iTRAQ Reagent 116, or iTRAQ Reagent 117.

9. Cap and seal the 2 mL protein lobind tube containing sample with parafilm carefully.

10. Vortex erectly at room temperature for 1 hour.

    - Do not vortex too vigorously.

11. Prepare a new 2 mL protein lobind tube. Spin down the labeled peptides and combine the four samples of labeled peptides to the new tube.

    - Do not combine the labeled peptides to one of the sample tube to ensure the amount of peptides of each sample in the combined tube is equal.

12. Dry the labeled peptides with centrifugal evaporator.

No need to make the sample too dry. Proceed to the next step when the labeled peptides become as a small drop of brown liquid.

ZipTip desalting

- For small-scale experiment, desalt the iTRAQ-labeled peptides directly.
- For large-scale experiment, we recommend performing strong cation exchange (SCX) chromatography after iTRAQ labeling. Dry every fraction of iTRAQ-labeled peptides with centrifugal evaporator and desalt each fraction individually.

1. Resuspend the dried iTRAQ-labeled peptides with $20-30\,\mu$L of 0.1% TFA.

2. Adjust the pH of each sample to about pH 2–3 with 10% TFA. Examine the pH with pH test strip.

   - Use the P20 pipetman for the following steps.

3. Pre-wet the ZipTip by aspirating 50% ACN/0.1% TFA into tip and dispense to waste on a paper towel. Repeat this step for 5 times.

4. Equilibrate the ZipTip by aspirating 0.1% TFA into tip and dispense to waste on paper towel. Repeat this step for 10 times.

5. Sample binding by aspirating and dispensing the sample for 20 cycles.

   - Be sure to aspirate completely to let $20\,\mu$L of sample pass through the membrane in ZipTip each cycle.

6. Wash the ZipTip by aspirating 0.1% TFA into tip and dispense to waste on paper towel. Repeat this step for 10 times.

7. Elute the peptides by aspirating 50% ACN/0.1% TFA into tip and dispense the eluent ($20\,\mu$L) into a new 2 mL protein lobind tube.

8. Aspirate and dispense the eluent in the tube for 10 cycles.

9. Dry the desalted peptides with centrifugal evaporator.
10. The sample is ready for LC-MS/MS analysis.

References

1. Wilkins MR, Pasquali C, Appel RD *et al.* (1996) From proteins to proteomes: Large scale protein identification by two-dimensional electrophoresis and amino acid analysis. *Biotechnology* (NY) **14(1):** 61–65.
2. Zhang Y, Fonslow BR, Shan B *et al.* (2013) Protein analysis by shotgun/bottom-up proteomics. *Chem. Rev.* **113:** 2343–2394.
3. Domon B and Aebersold R. (2006) Mass spectrometry and protein analysis. *Science* **312:** 212–217.
4. Yates JR. (1998) Mass spectrometry and the age of the proteome. *J. Mass Spectrom* **33:** 1–19.
5. Cox J and Mann M. (2011) Quantitative, high-resolution proteomics for data-driven systems biology. *Annu. Rev. Biochem.* **80:** 273–299.
6. Fenn JB, Mann M, Meng CK *et al.* (1989) Electrospray ionization for mass spectrometry of large biomolecules. *Science* **246:** 64–71.
7. Bantscheff M, Schirle M, Sweetman G *et al.* (2007) Quantitative mass spectrometry in proteomics: A critical review. *Anal. Bioanal. Chem.* **389:** 1017–1031.
8. Gerber SA, Rush J, Stemman O *et al.* (2003) Absolute quantification of proteins and phosphoproteins from cell lysates by tandem MS. *Proc. Natl. Acad. Sci. U S A* **100:** 6940–6945.
9. Mallick P and Kuster B. (2010) Proteomics: A pragmatic perspective. *Nat. Biotechnol.* **28:** 695–709.
10. Ong S-E, Blagoev B, Kratchmarova I *et al.* (2002) Stable isotope labeling by amino acids in cell culture, SILAC, as a simple and accurate approach to expression proteomics. *Mol. Cell. Proteomics* **1:** 376–386.
11. Gygi SP, Rist B, Gerber SA *et al.* (1999) Quantitative analysis of complex protein mixtures using isotope-coded affinity tags. *Nat. Biotechnol.* **17:** 994–999.
12. Hsu J-L, Huang S-Y, Chow N-H *et al.* (2003) Stable-isotope dimethyl labeling for quantitative proteomics. *Anal. Chem.* **75:** 6843–6852.
13. Schmidt A, Kellermann J and Lottspeich F. (2005) A novel strategy for quantitative proteomics using isotope-coded protein labels. *Proteomics* **5:** 4–15.
14. Thompson A, Schäfer J, Kuhn K *et al.* (2003) Tandem mass tags: A novel quantification strategy for comparative analysis of complex protein mixtures by MS/MS. *Anal. Chem.* **75:** 1895–1904.
15. Ross PL, Huang YN, Marchese JN *et al.* (2004) Multiplexed protein quantitation in Saccharomyces cerevisiae using amine-reactive isobaric tagging reagents. *Mol. Cell. Proteomics.* **3:** 1154–1169.
16. Mahoney DW, Therneau TM, Heppelmann CJ *et al.* (2011) Relative quantification: characterization of bias, variability and fold changes in mass spectrometry data from iTRAQ-labeled peptides. *J Proteome. Res.* **10:** 4325–4333.
17. Han C-L, Chien C-W, Chen W-C *et al.* (2008) A Multiplexed quantitative strategy for membrane proteomics opportunities for mining therapeutic targets for autosomal dominant polycystic kidney disease. *Mol. Cell. Proteomics* **7:** 1983–1997.

4. Phosphoproteome: Sample Preparation

Chia-Wei Hu and Hsueh-Fen Juan*

*Institute of Molecular and Cellular Biology,
National Taiwan University, Taipei, Taiwan*
*yukijuan@ntu.edu.tw

1. Introduction

Protein phosphorylation is one of the most important post-translational modifications in cells. It plays a crucial role in various biological processes including cell cycle, cell signaling, and metabolism. In eukaryote, it is believed that phosphorylation is involved in at least 40% of proteins.[1] Kinases and phosphatase are the two major enzymes that control the reversible phosphorylation/dephosphorylation mechanism using adenosine triphosphate (ATP) as its phosphate donor. In signal transduction, phosphorylation mediates signal attenuation and termination, whereas an aberrant regulation of protein phosphorylation often results in uncontrolled cell signaling, leading to a variety of diseases.[2,3] Therefore, the systematic study of phosphorylation events is an important determinant for understanding the regulation of cell physiology.

The efficient enrichment of phosphopeptides and high-resolution analyzer are highly required for the low presence of phosphorylated proteins in cell. Many strategies have been developed recently to selectively enrich phosphopeptides, including immobilized metal affinity chromatography (IMAC), titanium dioxide (TiO_2) enrichment, strong cation exchange chromatography, and antibody-based enrichment.[3,4] Among these approaches, TiO_2 is one of the most widely used methods in phosphoproteomic analysis.

On account of the Lewis acid–base character of TiO_2, the phosphate groups of peptides can efficiently bind to TiO_2 under acidic condition but desorbed under alkaline condition. However, the acidic peptides rich in glutamic and aspartic acids are often absorbed to the TiO_2, leading to the non-specific binding during phosphopeptide enrichment.[5] In 2008, hydroxyl acid-modified metal oxide chromatography (HAMMOC) was developed as a highly efficient method to enrich phosphopeptides.[6] An aliphatic hydroxy acid, such as lactic acid, is used to increase the specificity of phosphopeptide enrichment by competing with acidic non-phosphopeptides. Therefore, only phospho-peptides can successfully bind to the TiO_2 due to its high affinity.[6,7] In comparison with other strategies of phosphopeptide enrichment, HAMMOC is easy to perform, and the coverage on phosphopeptide identification can be improved.[7]

During the past few years, the detection of protein phosphorylation was extraordinarily improved due to the rapid development in mass spectrometry and proteomic technologies.[1,2] Based on the MS-based phosphoproteomics, complex protein samples are enzymatically digested into peptide mixtures, where the phosphopeptides are specifically enriched by affinity- or antibody-based methods. The peptide samples can be separated by nano-scale liquid chromatography and analyzed by mass spectrometry (nanoLC-MS/MS).[8] In this chapter, we describe a strategy to prepare samples for quantitative phosphoproteomic study by combining many previously published methods including phase transfer surfactant-aided trypsin digestion,[9] dimethyl label-ing,[10] and HAMMOC.[7] Many processes can be performed with StageTip.[11] This strategy provides a simple and efficient way for sample preparation before LC-MS/MS analysis.

2. Protein extraction by phase-transfer surfactant (PTS) buffer[9]

Materials:

- Sodium deoxycholate (SDC) (CAS NO. 302-95-4)
- Sodium N-lauroylsarcosinate (SLS) (CAS NO. 137-16-6)
- Protease inhibitor cocktails (100X)
- Ser/Thr phosphatase inhibitor cocktails (100X)
- Tyr phosphatase inhibitor cocktails (100X)

- 1 M Triethylammonium bicarbonate (TEAB) (CAS NO. 15715-58-9)/Tris-HCl (pH = 9.0)
- 1.5 mL microcentrofuge tube

Procedure:

I. Prepare 120 mM SDC in H_2O.

II. Prepare 120 mM SLS in H_2O.

III. Prepare cell lysis buffer (freshly prepared): 12 mM SDC, 12 mM SLS, 100 mM TEAB/Tris-HCl, and protease/phosphatase inhibitors (Table 1).

IV. Cell collection

> Method I: Wash cells with PBS and harvest cells in ice-cold lysis buffer by scrapping. Immediately store samples at −80°C until use.

> > **Hint**: Remove PBS completely before the addition of lysis buffer. The remaining PBS could dilute the lysis buffer.

> Method II: Wash cells with PBS. Harvest cells by trypsinization and centrifuge at 1,000 × g for 5 minutes. Cell pellet can be stored at −80°C until use. Resuspend the pellet in ice-cold lysis buffer before sonication.

V. Lyse cells using sonication with suitable amplitude and time (60% amplitude and 0.6 cycle for cultured cells using Sartorius LAB-SONIC M). Keep samples in ice-water bath to avoid the heating of samples. The lysate should become clear after sonication.

Table 1.   Recipe for cell lysis buffer.

PTS buffer for cell lysis	1 mL
120 mM SDC	100 μL
120 mM SLS	100 μL
1 M TEAB or Tris-HCl (pH 9.0)	100 μL
Protease inhibitor cocktails	10 μL
Ser/Thr phosphatase inhibitor cocktails	10 μL
Tyr phosphatase inhibitor cocktails	10 μL
MilliQ	670 μL

  VI. Remove cell debris by centrifuge at $16{,}000 \times g$ for 20 minutes, 4°C.

  VII. Transfer the supernatant to a new 1.5 mL tube. The protein extracts can be stored at -80°C until use.

3. Protein digestion and desalting

Materials:

- BCA protein assay kit (#23225, Thermo Scientific)
- Dithiothreitol (DTT) (CAS NO. 3483-12-3)
- Iadoacetamide (IAM) (CAs NO. 144-48-9)
- 1M TEAB/ammonium bicarbonate
- Lys-C endopeptidase (#129-02541, WAKO)
- Sequencing grade trypsin (#90305, Thermo Scientific)
- Ethyl acetate (EtAc) (CAS NO. 141-78-6)
- Empore^{TM} SDB-XC 6 mL extraction disc cartridge (#4340HD, 3M)
- Solution A: 0.1% TFA, 80% ACN

Procedure:

    I. Determine the protein concentration using BCA protein assay kit.

    II. Transfer suitable amount of proteins to a new tube. Adjust the protein concentration to 1–1.5 mg/mL using lysis buffer.

   III. Prepare 0.5 M DTT in 50 mM TEAB/ammonium bicarbonate (freshly prepared).

   IV. Protein reduction by adding 0.5 M DTT to a final concentration of 10 mM. Rotating for 30 minutes at room temperature.

    V. Prepare 1 M IAM in 50 mM TEAB/ammonium bicarbonate (freshly prepared).

   VI. Protein alkylation by adding 1 M IAM to a final concentration of 50 mM. Rotating for 30 minutes at room temperature in dark.

  VII. Add Lys-C to obtain an enzyme: protein ratio (mass: mass) of 1:100. Rotating for 3 hours at room temperature.

 VIII. Add trypsin to obtain an enzyme: protein ratio (mass: mass) of 1:50. Rotating for 16–18 hours at room temperature.

   IX. For detergent removal, add equal volume of EtAc to the peptide sample.

    X. Inactivate trypsin by adding 10% TFA to a final concentration of 0.5%.

   XI. Agitate for 2 minutes to completely mix EtAc and sample solution.

XII. Centrifuge at 15,700 g for 2 minutes at room temperature and remove the supernatant carefully.

> **Hint**: Do not remove the white disc between the organic and aqueous layers. It is acceptable to remove a few microliters of samples.

XIII. Vacuum dry the peptide samples.

XIV. Dissolve the peptides by adding $300-500\,\mu$L solution **A**. The precipitation from white disc cannot be dissolved.

XV. Peptide desalting using Empore^{TM} SDB-XC 6 mL cartridge according to the manufacturer's instructions.

> **Hint**: Load all the samples including precipitations to the desalting column to increase the recovery of peptides.

4. Dimethyl labeling of peptides[12]

Materials:

- 1 M TEAB
- 37% formaldehyde solution (CH_2O, CAS NO. 50-00-0)
- 20% formaldehyde-13C, D2 solution ($D^{13}CDO$, CAS NO. 63101-50-8)
- Cyanoborohydride ($NaBH_3CN$)

> **Hint**: Store $NaBH_3CN$ in a desiccator.

- 25% ammonia
- Formic acid

Procedure:

I. Buffer preparation (freshly prepared):

   i. 100 mM TEAB
   ii. 4% CH_2O
   iii. 4% $^{13}CD_2O$
   iv. 0.6 M $NaBH_3CN$

II. Dissolve 100 μg peptides by 100 μL 100 mM TEAB (100 μL can be used for a maximum of 100 μg peptides).

The reagent amount used in the following steps are designed for $100\,\mu L$ samples:

III. Add $4\,\mu L$ of CH_2O (light) or $^{13}CD_2O$ (heavy) to different groups of samples.
IV. Add $4\,\mu L$ of $NaBH_3CN$ to each group of sample as a catalyst for the labeling reaction.
V. Agitate for 1 minute to completely mix the reagents and samples.
VI. Incubation by rotating at room temperature for 60 minutes.
VII. Buffer preparation:

    i. 1% ammonia
    ii. 10% formic acid

VIII. Spin down and put the samples on ice.
IX. Add $16\,\mu L$ of 1% ammonium to consume the remaining labeling reagents.
X. Agitate samples for 1 minute and then spin down the samples.
XI. Add $20\,\mu L$ of 10% formic acid to further stop the labeling reaction.
XII. Agitate samples for 1 minute and then spin down the samples.
XIII. Transfer the differentially labeled samples to a new $1.5\,mL$ tube.
XIV. Agitate for 1 minute to completely mix the samples.
XV. Desalting using Empore^{TM} SDB-XC $6\,mL$ cartridge.

5. Preparation and preservation of TiO_2

Materials:

- Titansphere TiO_2 (#5020-75010, GL science)

Procedure:

I. The newly purchased titansphere TiO_2 should be heat-treated at $500°C$ for 4 hours for complete dehydration and decomposition of polymers. The color of TiO_2 would change from pale yellow to white after heating treatment.
II. Store TiO_2 in a desiccator.

6. Preparation of StageTips and tubes[11]

Materials:

- Empore C8 disc (#22214, 3M)

- P10 pipette tip
- 20 gauge blunt-point needle (#90520, Hamilton)
- Plunger for 20 gauge needle (#13205, Hamilton)
- Emphore SDB-XC disk (#2340, 3M)
- 16 gauge blunt-point needle (#90516, Hamilton)
- Plunger for 16 gauge needle (#1122-01, Hamilton)
- P200 pipette tip
- 1.5 mL microcentrifuge tube

Procedure:

I. Use 20G needle to cut the C8 disc, and then push the disc into the P10 tip by plunger (leave around 1 mm distance from the end of the tip) (Figs. 1(a)–1(c)).

II. Use 16 gauge needle to cut the SDB-XC disc, and then push the membrane into the P200 tip by plunger (Figs. 1(d)–1(e)).

III. Bore one large and one small hole on the cap of 1.5 mL tube. The large hole is for inserting StageTip. The small hole is for ventilation.

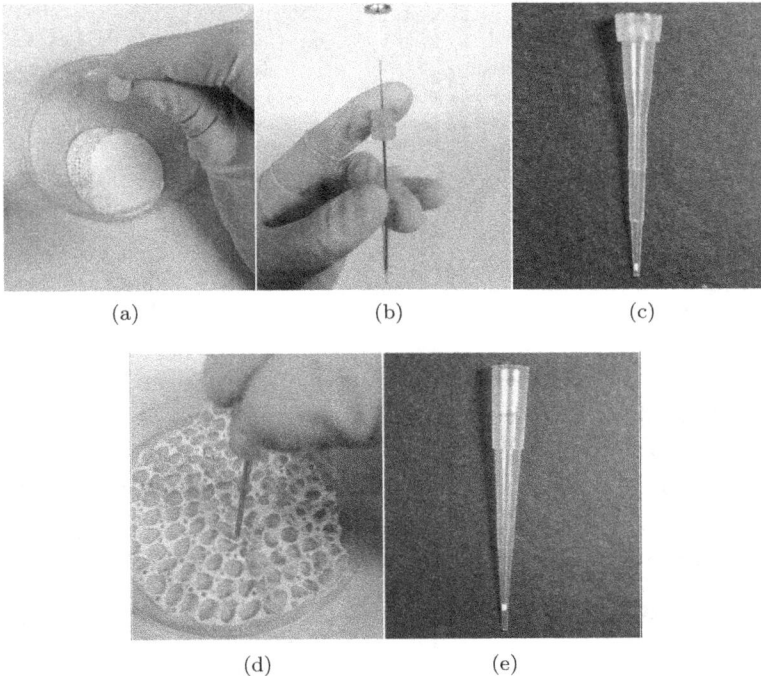

(a) (b) (c)

(d) (e)

Figure 1. Assembly of C8 and SDB-XC StageTips. (a) Cut suitable size of C8 disc by using 20G needle. (b) Push the disc into the P10 tip. (c) The C8 StageTip. (d) Cut suitable size of SDB-XC disc by using 16G needle. (e) The small SDB-XC StageTip.

7. Hydroxy-acid modified metal oxide chtomatography (HAMMOC)[3]

Materials:

- Lactic acid (LA) (CAS NO. 50-21-5)
- Piperidine (CAS NO. 110-89-4)
- Trifluoroacetic acid (TFA) (CAS NO. 76-05-1)
- Phosphoric acid (CAS NO. 7664-38-2)
- Acetonitrile (ACN), HPLC grade
- 1.5 mL microcentrifuge tube (#0030 120.086, Eppendorf)
- Pipette tips
- C8 StageTips in 1.5 mL microcentrifuge tube
- SDB-XC StageTips in 1.5 mL microcentrifuge tube
- Methanol (MeOH)

Procedure:

I. Buffer preparation:
   Solution A: 0.1% TFA, 80% ACN
   Solution B: 0.1% TFA, 5% ACN
   Solution C: 300 mg/mL LA
   Solution D: 0.5% piperidine

TiO_2 packing

II. Take TiO_2 from the desiccator and weight the TiO_2 spheres in a 1.5 mL tube (0.5 mg TiO_2/tip).

   Hint: The weighted TiO_2 should be higher than the required amount. For example, if you need 2 mg TiO_2, prepare 2.5–3 mg TiO_2.

III. Add MeOH to the TiO_2 tube (40 μL MeOH/mg TiO_2. Ex. for 1.5 mg TiO_2, add 60 μL MeOH). Put a tiny stir bar into the tube. Wash the stir bar with MeOH before use.

IV. Allow TiO_2 and MeOH to be continuously mixed on a plate stirrer. Transfer 20 μl TiO_2 (0.5 mg) solution to each C8 StageTip.

V. Centrifuge at 500 × g for 3–5 minutes to remove MeOH.

Phosphopeptide enrichment

VI. **Precondition** the TiO_2 tips by loading 20 μL solution **B**, centrifuge at 1,500 × g for 2 minutes.

> **Hint**: Examine the tip after every centrifuge step to make sure there is no solution remained in the tip.

VII. **Equilibrate** the tips by loading $20\,\mu L$ solution **C**, centrifuge at $1{,}500 \times g$ for 2 minutes.

VIII. **Load** $50\,\mu L$ of sample, centrifuge at $1{,}000 \times g$ for 2 minutes. Repeat the step until all the peptide samples are loaded into the StageTip.

> **Hint**: the loading capacity for each C8 StageTip is $100\,\mu g$ peptides.

IX. **Wash** tips by loading $50\,\mu L$ solution **C**, centrifuge at $1{,}500 \times g$ for 2 minutes.

X. **Wash** tips by loading $50\,\mu L$ solution **B**, centrifuge at $1{,}500 \times g$ for 2 minutes. Prepare new $1.5\,mL$ microcentrifuge tubes with boreholes for elution (one tube for each StageTip).

XI. Transfer the StageTip to the new $1.5\,mL$ tube.

XII. **Elute** phosphopeptides by loading $50\,\mu L$ solution D (0.5%), centrifuge at $1{,}000 \times g$ for 2 minutes.

XIII. Repeat step XII. Acidify the eluted peptides by adding 10% TFA to a final concentration of 0.5% and examine the pH value by test paper.

> **Hint 1**: Peptides are more stable in the environment of low pH compared to alkaline condition.
> **Hint 2**: Different elution buffers such as ammonium bicarbonate might elute different phosphopeptides.

Desalting

XIV. Each TiO_2 tip requires one SDB-XC StageTips for desalting.

XV. **Precondition** the SDB-XC tips by loading $20\,\mu L$ solution **B**, centrifuge at $1{,}000\,g$ for 1 minute.

> **Hint**: Be careful about the height of StageTip upon the tube. Do not let the tip touch the cover of centrifuge machine.

XVI. **Equilibrate** the tips by loading $20\,\mu L$ solution **A**, $1{,}000 \times g$ for 1 minute.

XVII. **Load** the sample to the StageTip, centrifuge at $1{,}000 \times g$ for 1 minute. Repeat the step until all the peptide samples are loaded into the StageTip.

XVIII. **Transfer** the tip to the new tube.

XIX. **Elute** peptides by adding $20\,\mu L$ solution **B**, $1,000 \times g$ for 1 minute.

XX. Transfer the eluted samples to a new 1.5 mL tube. Sample can be stored at $20°C$ for one month.

Hint: Post-fractionation of phosphopeptide samples can be performed to increase the number of identified phosphopeptides

XXI. Vacuum dry the samples.

XXII. Dissolve peptides by 0.1 % TFA and subject to LC-MS/MS.

References

1. Junger MA and Aebersold R. (2014) Mass spectrometry-driven phosphoproteomics: Patterning the systems biology mosaic. *Wiley Interdiscip. Rev. Dev. Biol.* **3**: 83–112.
2. Macek B, Mann M, and Olsen JV. (2009) Global and site-specific quantitative phosphoproteomics: Principles and applications. *Annu. Rev. Pharmacol. Toxicol.* **49**: 199–221.
3. Olsen JV, Blagoev B, Gnad F *et al.* (2006) Global, *in vivo*, and site-specific phosphorylation dynamics in signaling networks. *Cell* **127**: 635–648.
4. Amoresano A, Cirulli C, Monti G *et al.* (2009) The analysis of phosphoproteomes by selective labelling and advanced mass spectrometric techniques. *Methods Mol. Biol.* **527**: 173–190.
5. Bodenmiller B, Mueller LN, Mueller M *et al.* (2007) Reproducible isolation of distinct, overlapping segments of the phosphoproteome. *Nat. Methods* **4**: 231–237.
6. Ku W-C, Sugiyama N and Ishihama Y. (2012) Large-scale protein phosphorylation analysis by mass spectrometry-based phosphoproteomics. *Protein Kinase Technologies* **68**: 35–46.
7. Sugiyama N, Masuda T, Shinoda K *et al.* (2007) Phosphopeptide enrichment by aliphatic hydroxy acid-modified metal oxide chromatography for nano-LC-MS/MS in proteomics applications. *Mol. Cell. Proteomics* **6**: 1103–1109.
8. Bantscheff M, Schirle M, Sweetman G *et al.* (2007) Quantitative mass spectrometry in proteomics: a critical review. *Anal. Bioanal. Chem.* **389**: 1017–1031.
9. Masuda T, Tomita M, and Ishihama Y. (2008) Phase transfer surfactant-aided trypsin digestion for membrane proteome analysis. *J. Proteome. Res.* **7**: 731–740.
10. Boersema PJ, Raijmakers R, Lemeer S *et al.* (2009) Multiplex peptide stable isotope dimethyl labeling for quantitative proteomics. *Nat. Protoc.* **4**: 484–494.
11. Rappsilber J, Mann M and Ishihama Y (2007) Protocol for micro-purification, enrichment, pre-fractionation and storage of peptides for proteomics using StageTips. *Nat. Protoc.* **2**: 1896–1906.
12. Wilhelm S, Carter C, Lynch M *et al.* (2006) Discovery and development of sorafenib: a multikinase inhibitor for treating cancer. *Nat. Rev. Drug. Discov.* **5**: 835–844.
13. Kyono Y, Sugiyama N, Imami K *et al.* (2008) Successive and selective release of phosphorylated peptides captured by hydroxy acid-modified metal oxide chromatography. *J. Proteome Res.* **7**: 4585–4593.

5. Transcriptomic Data Analysis: RNA-Seq Analysis Using Galaxy

Chia-Lang Hsu*,‡ and Chantal Hoi Yin Cheung†

*Department of Life Sciences, National Taiwan University,
Taipei, Taiwan
†Institute of Molecular and Cellular Biology,
National Taiwan University, Taipei, Taiwan
‡auymle@gmail.com

Introduction

The global characterization and profiling of transcriptome are critical steps for unveiling RNA which plays a multifaceted role in numerous biological processes. In cancer research, transcriptome analyses have been utilized to identify aberrant transcripts associated with specific pathogenetic mechanisms and a group of genes, namely the gene signature for distinguishing cancer subtypes or predicting prognosis of cancer patients. With the evolution of next-generation sequencing (NGS) technologies, traditional methods, such as expressed sequence tag (EST) and gene expression microarray, have been complemented by massively parallel RNA sequencing (RNA-seq). RNA-seq offers excellent ability to unbiasedly quantify transcript expression in a single assay. In addition, RNA-seq has the potential ability to detect new genes and transcript isoforms, genetic variants (e.g., single nucleotide variants, insertions, and deletions), and gene fusions. RNA-seq opens up an entirely new age of transcriptome analysis in cancer research.

RNA-seq is being increasingly used, in part driven by the decreasing cost of sequencing. Nevertheless, the manipulation and analysis of the massive amounts of data generated by RNA-seq is a challenge for scientists who lack the background knowledge for programming. Galaxy is a web-based platform that makes computational biomedical research accessible to the users without prior programming knowledge.[1–3] It collects various genomics and NGS analysis tools, and provides graphical interfaces to specify kinds of data, with a step-by-step historical view for the operating procedures. Another goal of Galaxy is to make computational research reproducible and transparent. To make analysis reproducible and transparent, Galaxy automatically generates the metadata for each analysis step, and the users can share or publish their datasets, analysis history or workflow via Galaxy. Because Galaxy is an open source and browser-based platform, it can be accessed by anyone and is free of charge. You can use the public server or install Galaxy on your personal computer. Galaxy is able to fill the gap between computer science and biology.

In this protocol, we will illustrate how to analyze RNA-seq dataset and identify the differentially expressed genes between two biological conditions using Galaxy. RNA-seq analysis contains three general steps: (i) read mapping, (ii) expression quantification, and (iii) differential expression analysis. Since each step can be supported by many tools, we will demonstrate more than one tool for each step in order to cover more analysis tools for the users. Figure 1 highlights the analysis tools used in this protocol. It is important to note that each tool only accepts input dataset with the appropriate file format and generates the file with a specific datatype. The common file formats in RNA-seq analysis are summarized in Table 1.

Materials

Hardware requirement

Galaxy is a Web-based tool, a computer with a modern web browser that supports JavaScript and HTML5 is required for the interaction of Galaxy interface.

Input data

In this protocol, we will use RNA-seq datasets of three pancreatic cancer cell lines (MiaPaCa2, PANC1, and HPAC) which can be found in the Galaxy

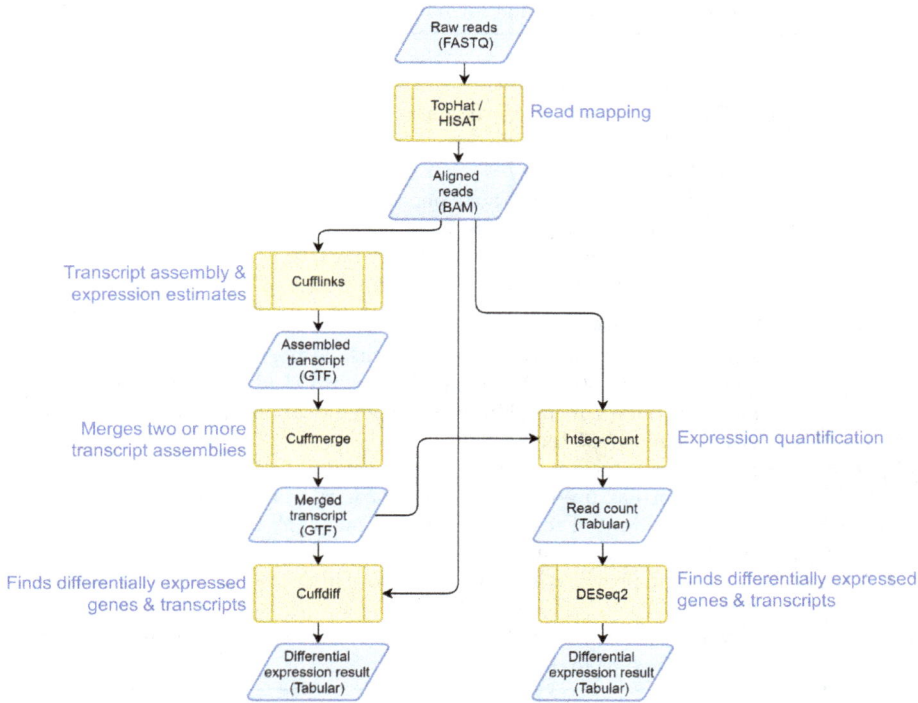

Figure 1.   The overview of RNA-seq analysis protocol. This protocol uses two different statistic tools for identification of differentially expressed genes.

library. These are paired-end datasets that we will use as an example to illustrate the results obtained from Galaxy.

Procedure

Access galaxy

First, start the web browser and go to the Galaxy Project's public server page, https://usegalaxy.org/. This is the main public Galaxy server maintained by the Galaxy Project team. The first interface you will encounter is the "Analyze Data" which is divided into four main parts (Fig. 2). The masthead contains the help, account setting, and shared data pages. On the left panel is the Tool menu, listing the available tools and software in Galaxy. The users' analysis histories are listed on the right panel. The middle panel is the tool interface which allows the users to select input datasets, to configure tool parameters, and to view the analysis results.

Table 1.   Common file formats for RNA-seq analysis

Format	Description
FASTQ	The FASTQ format stores sequences and qualities scores in a single file. FASTQ file uses four lines per sequence: • Line 1: begins with a "@" character and is followed by a sequence identifier with/without a sequence description • Line 2: the raw sequence letters • Line 3: begins with a "+" symbol • Line 4: encodes the quality values for the sequence in Line 2
SAM/BAM	SAM format data is the output from aligners that read FASTQ files and assign the sequences to a position with respect to a known reference genome. BAM stores the same data as SAM file in a compressed, indexed, binary form. These two formats can easily convert to each other using SAM tools
GTF/GFF	The gene transfer format (GTF) and general feature format (GFF) are the file format used to describe information about gene structure. GTF is a tab-delimited text format based on GFF, but contains some additional conventions specific to gene information.
BED	BED format provides a flexible way to define the data lines that are displayed in an annotation track
WIG	The wiggle (WIG) format is for display of dense, continuous data such as the coverage of mapped reads.

Create a galaxy account

As a registered user on the Galaxy main server, Galaxy will allow all the work to persist between sessions.

1. In the masthead at the top of page, click the "Register" link under the "User".
2. Provide your information to Galaxy by entering your valid email address, password, and confirm your password.
3. Your Galaxy account will be created by clicking the submit button.

Upload sequence data to galaxy

You can import data from your local storage either from a website or FTP site, you can even retrieve data from different databases, such as EBI SRA and UCSC via "Upload file" function.

1. In the Tools panel, expand "Get Data" and click "Upload file".
2. Drag and drop files into the pop-up window, or click "Choose local file" to select the files from your local storage. If your files are from a website, click "Paste/Fetch data" and paste the URL into the text-entry box.

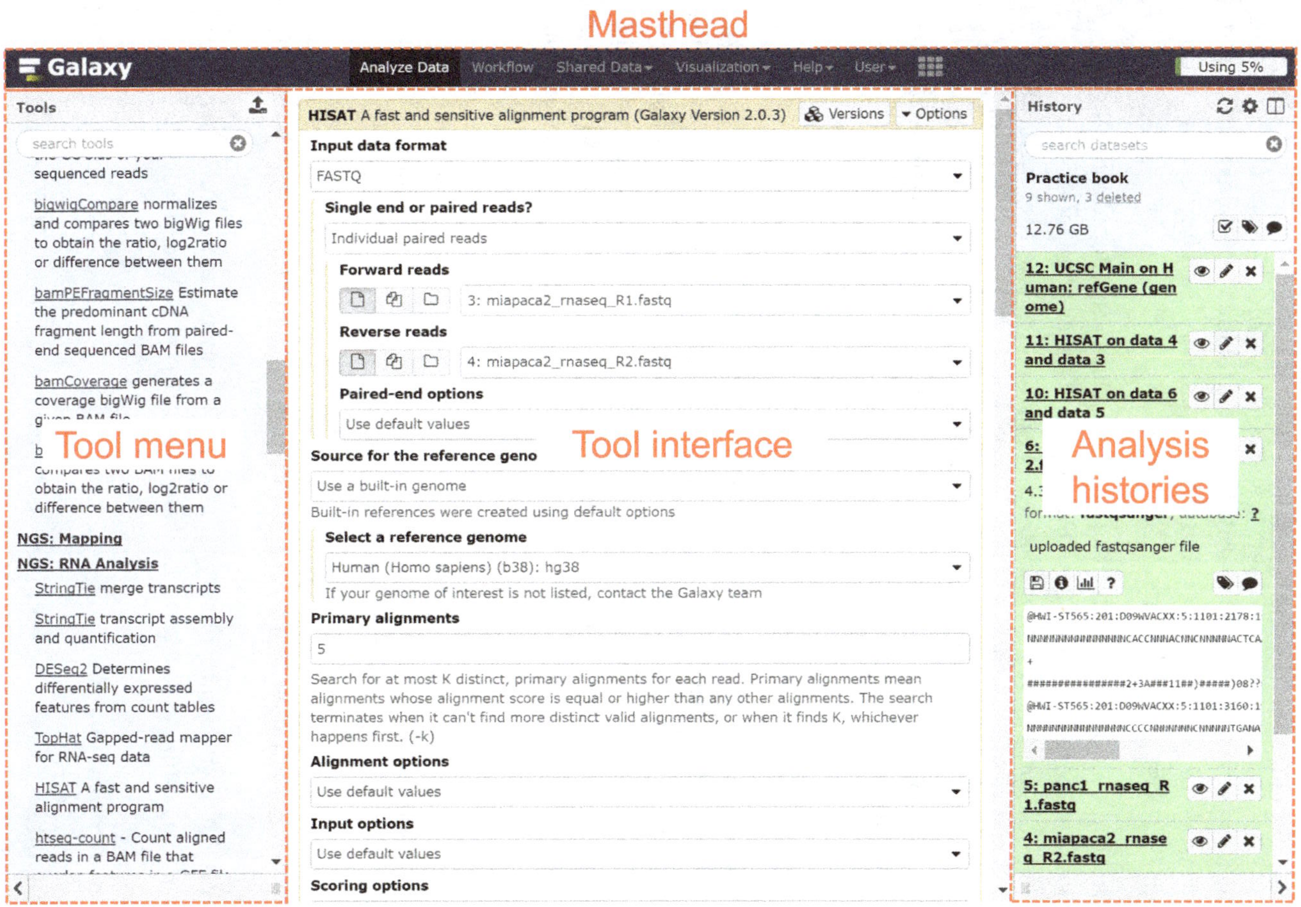

Figure 2. The Galaxy analysis interface. The interface consists of four main parts: the masthead at the top, the tool menu on the left, the analysis history on the right, and the tool interface in the center.

3. Click "Start" to upload the files to Galaxy. As upload processing is completed, a history item will be created on the left panel. Then, click "Close" to close the pop-up window.

You can also obtain data from Data Libraries provided by Galaxy or shared by other users. In this protocol, the RNA-seq datasets used for demonstration are from Galaxy's data libraries.

1. Click "Data Libraries" under the "Shared Data" in the masthead.
2. Choose the data library name "Pancreatic Cancer Cell Lines".
3. Check the boxes of "hpac_rnaseq_R1.fastq", "hpac_rnaseq_R1.fastq", "miapaca2_rnaseq_R1.fastq", "miapaca2_rnaseq_R2.fastq", "panc1_rnas eq_R1.fastq", and "panc1_rnaseq_R2.fastq", and then click "to History".
4. Click "Analyze Data" at the masthead, and the imported data will be listed in the History panel.

View data and edit attributes

The datasets listed in the History can be viewed, edited, or deleted (Fig. 3). By clicking the eye icon in the history panel, the content of the dataset will be illustrated in the main Galaxy panel. Also, you can change the name of a dataset displayed in the History panel by clicking the pencil icon to modify the "Name" in the "Attributes" tab and then click "Save". If you would like to save the dataset on your computer, click on the name of the dataset, you will find a slightly larger preview version in a new window and then click the disk icon.

It is important to note that many tools only accept input datasets with an appropriate datatype assigned. Although datatype can be automatically detected by Galaxy, please check the datatype after uploading a file. The datatype can be altered by the following method:

1. Click the pencil icon inside of a dataset box, then click the "Datatype" tab.
2. Select the appropriate datatype in the pulldown menu, and then click "Save" to complete the modification of datatype of this dataset.

Upload annotation file

The annotation file can be obtained and saved as GTF or GFF format from genome databases, such as NCBI and Ensemb. Here, we demonstrated two different ways to get the annotation files:

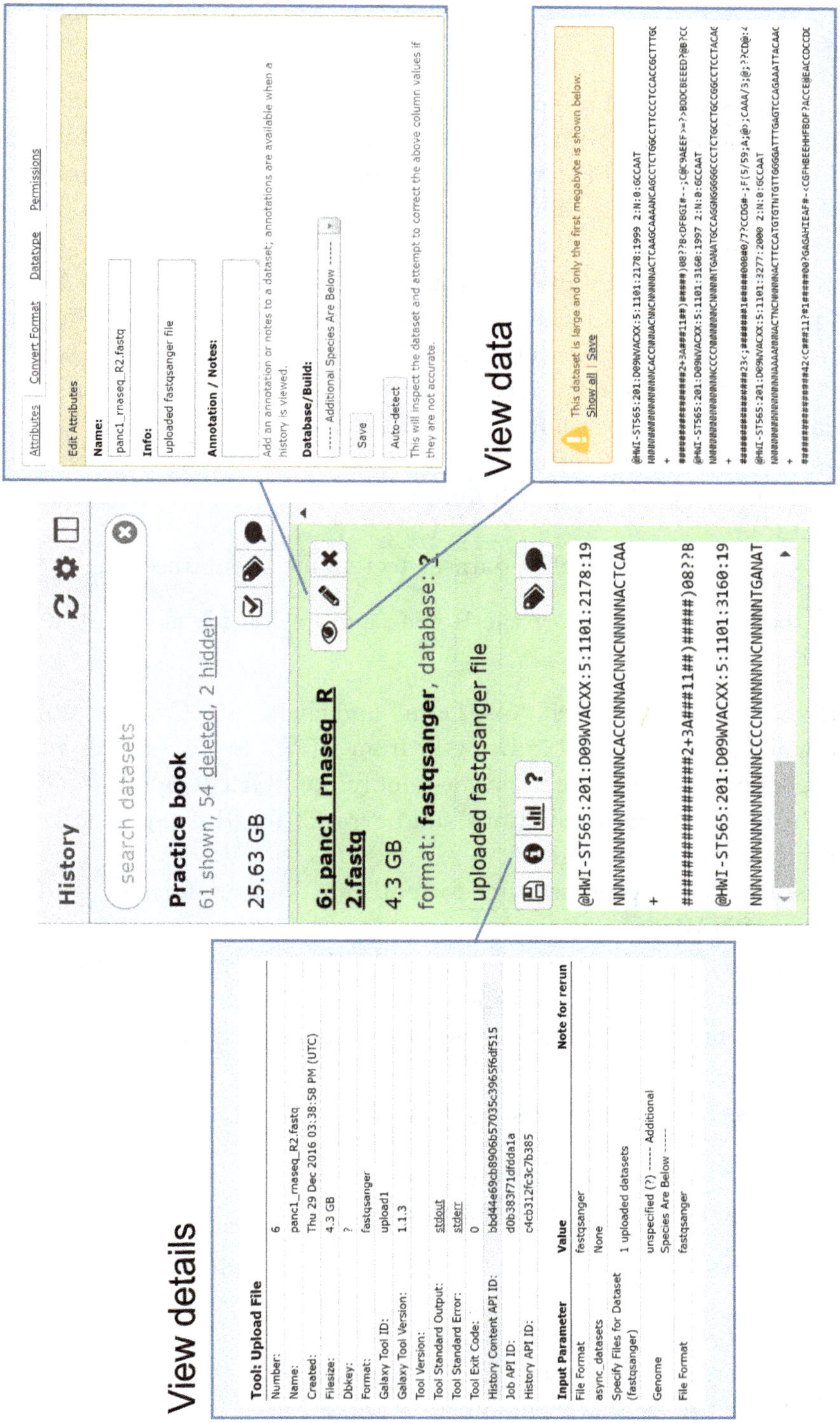

Edit attributes

View data

View details

Figure 3. View and edit a dataset in the History panel.

The first way to achieve the annotation file is from the GENCODE project.[4]

1. Go to the GENCODE official website (https://www.gencodegenes.org/).
2. You can find all the available files in the page of human current release, and download the GTF or GFF file or copy the link address directly.
3. Back to the Galaxy website. In the Tools panel, expand "Get Data" and click "Upload file".
4. If the files have been saved on your computer, drag and drop files into the pop-up window, or click "Choose local file" to choose the files from your computer. Alternatively, click "Paste/Fetch data", then paste the URL, e.g., ftp://ftp.sanger.ac.uk/pub/gencode/Gencode_human/release_25/gencode.v25.basic.annotation.gff3.gz, into the text-entry box, and then click the "Start" button. In this case, the original file is compressed as gz format, and Galaxy will automatically decompress this file as gff3 format.
5. Click the pencil icon of that dataset to rename the dataset.

The second way to import the RefSeq annotation file is from the UCSC interface.

1. In the Tools panel, expand "Get Data" and click "UCSC main" link. This tool will open up the Table Browser from UCSC in the Galaxy window.
2. Set "genome" to "human" and "assembly" to "GRCh38/hg38", "group" to "Genes and Gene Predictions" and "track" to "RefSeq Genes".
3. Select "genome" for "region".
4. Choose "GTF — gene transfer format" for "output format" and check the box of "Galaxy".
5. Click "get output" button and then click "Send query to Galaxy" button.
6. Imported RefSeq annotation file will be listed in the History panel, and click the pencil icon of that dataset to rename the dataset.

Read quality assessment using FastQC

Before starting to do RNA-seq analysis, perform the FastQC to check for any unusual qualities for sequence reads.

1. In the Tools panel, expand "NGS: QC and manipulation" and click "FastQC".
2. Under "Short read data from your current history", select a single or multiple fastq files for assessment of read quality.
3. Click "Execute" to start the job.

4. FastQC will generate a basic text and a HTML output file that contain all of the results, including basic statistics, quality score per base and per sequence, sequence content and so on.

Trimming sequences (optional)

Based on the results of quality control analysis, you may want to remove the low quality segments or reads for downstream analysis. For example, the figure of per base sequence quality generated by FastQC reveals that the quality scores of 3′ end of reads are very low (Fig. 4), use "Trim sequence" tool to remove the low quality regions as follows:

1. Expand "NGS: QC and manipulation" in the Tools panel and click "Trim sequence".
2. Under "Library to clip", click "Multiple datasets" and select all fastq files for trimming.
3. If you would like to remove the 10 bases from 3′ end of a read length of 100 bps. Set "First base to keep" to "1" and "Last base to keep" to "90".
4. Click "Execute" to start the job.

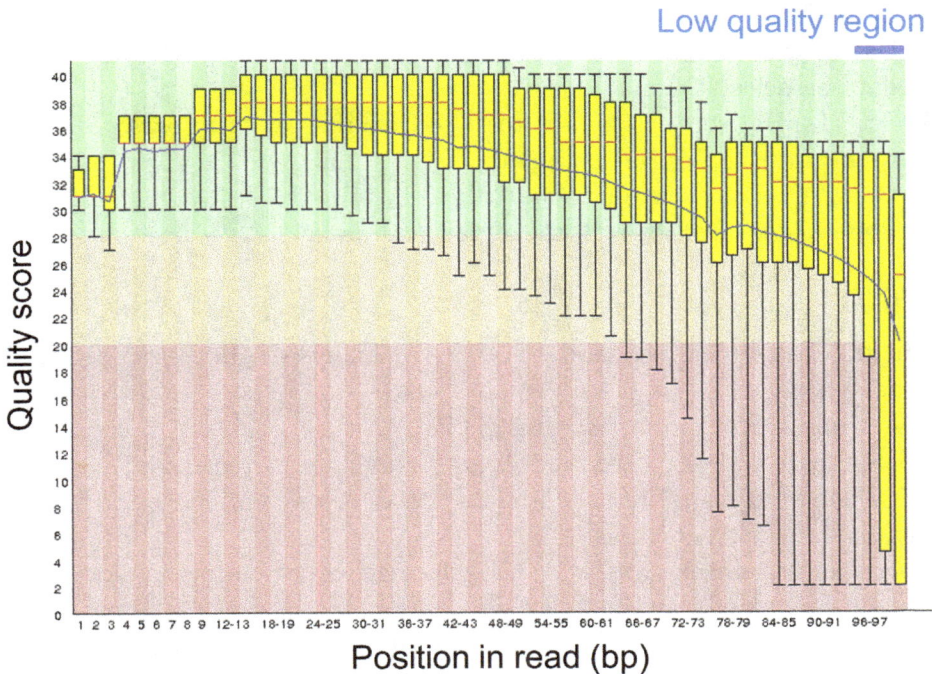

Figure 4. Distribution of per base sequence quality generated by FastQC.

Mapping reads using TopHat or HISAT

In this step, we demonstrated how to use two alignment algorithms, TopHat[5] and HISAT,[6] to generate read alignment data (BAM) for downstream analysis.

TopHat:

1. In the Tools panel, expand "NGS: RNA Analysis" and select "TopHat" tool.
2. Set "Is this single-end or paired-end data?" to "Pair-end (as individual datasets)".
3. Set "Single end or paired reads?" to "Individual paired reads".
4. Set "RNA-Seq FASTQ file, forward reads" to the FASTQ dataset corresponding to the forward reads, i.e., name contains "_R1".
5. Set "RNA-Seq FASTQ file, reverse reads" to the FASTQ dataset corresponding to the reverse reads, i.e., name contains "_R2".
6. Set "Mean Inner Distance between Mate Pairs" and "Std. Dev for Distance between Mate Pairs" based on the protocol of library construction. For example, for paired end runs with fragments selected at 300 bp, where each end is 50 bp, you should set "Mean Inner Distance between Mate Pairs" to 200.
7. Use a built-in genome and select the proper reference genome. In our dataset, select "Human: hg38".
8. Set "TopHat settings to use" to "Use Defaults" or "Full parameter list" to adjust each TopHat parameter.
9. Click "Execute" button to start the job.
10. Repeat steps 1–9 for each set of pair-end data.

HISAT:

1. In the Tools panel, expand "NGS:RNA Analysis" and select "HISAT" tool.
2. Set "Input data format" to "FASTQ".
3. Set "Single end or paired reads?" to "Individual paired reads".
4. Set "Forward reads" to the FASTQ dataset corresponding to the forward reads, i.e., name contains "_R1".
5. Set "Reverse reads" to the FASTQ dataset corresponding to the reverse reads, i.e., name contains "_R2".
6. Use a built-in genome and select the proper reference genome. In our dataset, select "Human: hg38".

7. Click "Execute" button to start the job.
8. Repeat steps 1–7 for each set of pair-end data.

Assemble transcripts using Cufflinks

The read alignment files produced by TopHat or HISAT are provided to Cufflinks to generate a transcriptome assembly and quantify the expression level of each transcript fragment for each sample.[7]

1. In the Tools panel, expand "NGS:RNA Analysis" and click "Cufflinks".
2. In the "SAM or BAM file of aligned RNA-Seq reads", use "Multiple datasets" and select the datasets produced by TopHat (accept_hit) or HISAT.
3. Click "Execute" button to run the analysis.

Merge assembled transcripts and map to a reference annotation using Cuffmerge

Because Cufflinks generates assembled transcripts for each sample, it is required to merge them together using Cuffmerge. In addition, when a reference genome annotation is available, Cuffmerge can integrate reference transcripts into the merged assembly.

1. In the Tools panel, click "NGS:RNA Analysis" to expand tool menu.
2. Click "Cuffmerges".
3. Set "GTF files produced by Cufflinks" to the "assembled transcripts" GTF file.
4. Click "Insert Additional GTF inputs" button for each additional "assembled transcripts" dataset.
5. Choose "Yes" for "Use Reference Annotation" and select the GTF annotation files exported from UCSC.
6. Click "Execute" button to run the analysis.

Differential expression analysis with Cuffdiff

Cuffdiff calculates expression in two or more samples and tests the statistical significance of each observed change in their expression.[7]

1. In the Tools menu, expand "NGS:RNA Analysis" and click "Cuffdiff".
2. Set "Generate SQLite" to "Yes". This file is used for visualizing the results of differential expression analysis.

3. Set "Transcripts" to the "merged transcripts" generated by Cuffmerge.
4. In "Condition" regions, set "Name" to the proper label (e.g., MiaPaCa2) and set "Replicates" input to the read-aligned SMA/BAM file generated by TopHat or HISAT for each of the RNA-seq conditions. If your experiment has more than two conditions, include additional conditions by clicking "Insert Condition".
5. If using only one sample per condition, "Dispersion estimation method" must be "blind".
6. Set "False Discovery Rate" to "0.05" or less.
7. Click "Execute" to run the analysis.

Visualize Cuffdiff output via CummeRbund

CummeRbund can visualize and integrate all of the data generated by a Cuffdiff analysis which helps the users to explore their expression data.

1. In the Tools panel, expand "NGS: RNA Analysis" and click "cummeR-bund".
2. Set "Select backend database (sqlite)" input to the SQLite file generated by Cuffdiff.
3. Click "Insert Plots" and set "Plot type" of your interest.
4. Click "Execute" to generate the plot.

Select differentially expressed genes

Finally, we can use Filter function provided by Galaxy to selected differentially expressed genes.

1. In the Tools panel, expand "Filter and Sort" and click "Filter".
2. Set "Filter" to the file "gene differential expression testing" generated by Cuffdiff.
3. Set "With following condition" to "c14 == 'yes'" where c14 denotes the 14th columns of the table.
4. Set "Number of header lines to skip" to "0".
5. Click "Execute" button.
6. A file will be generated in history panel and click the eye icon to view the dataset. In addition, this file can be used in further analysis, such as functional or pathway analyses.

Differential expression analysis using DESeq2

Besides Cuffdiff, Galaxy provides several tools for identification of differentially expressed genes or transcripts. In this section, we demonstrate how to use htseq-count[8] to quantify expression and DESeq2[9] for differential expression analysis. For example, we would like to compare the expression between PANC1 and HPAC cell lines.

1. Expand "NGS: RNA Analysis" in the Tools panel, and click "htseq-count" tool to quantify the expression of each gene or transcript.
2. Set "Aligned SAM/BAM File" input to files generated by TopHat (accept_hit) or HISAT.
3. Set "GFF File" input to file generated by Cuffmerge or annotation file obtained from other databases.
4. Choose "Intersection (nonempty)" for "Mode".
5. Select "Yes" for "Stranded".
6. Set "Feature type" to "exon" for transcript-level expression or to "gene" for gene-level expression.
7. Click "Execute" to start the job. As the job is complete, the expression datasets quantified by read count will be generated.
8. Click "DESeq2" tool, and choose "cell_line" for "Specify a factor name".
9. Set "Specify a factor level" to "PANC1" and "HPAC", respectively, and set "Count file(s)" input to the corresponding read count files produced by htseq-count.
10. Set "Choice of Input data" to "Count data".
11. Set "Visualising the analysis results" and "Output normalized counts table" to "Yes".
12. Click "Execute" to start the differential expression analysis. DESeq2 generates a tabular file containing the different columns and optional visualized results as PDF.
13. Expand "Filter and Sort" and click "Filter" tool to select the significantly expressed transcripts or genes.
14. Set "Filter" input to the dataset generated by DESeq2.
15. Set "With following condition" to "c7 < 0.05" where the 7th column of this dataset is the adjusted p-value. These files can be used in further analysis, such functional or pathway analyses.

Notes

Galaxy not only supports RNA-seq analysis but also other types of NGS datasets, such as ChIP-seq and exome-seq. In addition to this protocol, users interested in the analysis of other types of NGS dataset are directed to follow the tutorial available at Refs. 10, 11.

Acknowledgments

We specially thank the Galaxy team for their efforts on development and maintenance of Galaxy.

References

1. Giardine B, Riemer C, Hardison RC *et al.* (2005) Galaxy: A platform for interactive large-scale genome analysis. *Genome Res.* **15**(10): 1451–1455.
2. Goecks J, Nekrutenko A, Taylor J *et al.* (2010) Galaxy: A comprehensive approach for supporting accessible, reproducible, and transparent computational research in the life sciences. *Genome Biol.* **11**(8): R86.
3. Afgan E, Baker D, van den Beek M *et al.* (2016) The Galaxy platform for accessible, reproducible and collaborative biomedical analyses: 2016 update. *Nucleic Acids Res.* **44**(W1): W3–W10.
4. Harrow J, Frankish A, Gonzalez JM *et al.* (2012) GENCODE: The reference human genome annotation for The ENCODE Project. *Genome Res.* **22**(9): 1760–1774.
5. Kim D, Pertea G, Trapnell C *et al.* (2013) TopHat2: Accurate alignment of transcriptomes in the presence of insertions, deletions and gene fusions. *Genome Biol.* **14**(4): R36.
6. Kim D, Langmead B, and Salzberg SL. (2015) HISAT: A fast spliced aligner with low memory requirements. *Nat. Meth.* **12**(4): 357–360.
7. Trapnell C, Williams BA, Pertea G *et al.* (2010) Transcript assembly and quantification by RNA-seq reveals unannotated transcripts and isoform switching during cell differentiation. *Nat. Biotechnol.* **28**(5): 511–515.
8. Anders S, Pyl PT, and Huber W. (2015) HTSeq — A Python framework to work with high-throughput sequencing data. *Bioinformatics.* **31**(2): 166–169.
9. Love MI, Huber W, and Anders S. (2014) Moderated estimation of fold change and dispersion for RNA-seq data with DESeq2. *Genome Biol.* **15**(12): 550.
10. Blankenberg D and Hillman-Jackson J. (2014) Analysis of next-generation sequencing data using Galaxy. *Methods Mol Biol.* **1150**: 21–43.
11. Blankenberg D, Von Kuster G, Coraor N *et al.* (2010) Galaxy: A web-based genome analysis tool for experimentalists. *Curr. Protoc. Mol. Biol.* **Chapter 19:** Unit 19 1–21.

6. Proteomic Data Analysis: Functional Enrichment

Hsin-Yi Chang and Hsueh-Fen Juan*

Institute of Molecular and Cellular Biology, National Taiwan University, Taipei, Taiwan
*yukijuan@ntu.edu.tw

1. Introduction

High-throughput protein identification and quantification analysis based on mass spectrometry are evolving at a rapid pace. The state-of-art mass spectrometry provides platform to identify complicated proteome with high sensitivity at a relatively low cost and high reproducibility.[1,2] Data acquired from mass spectrometry need to be processed, managed, visualized, and analyzed in advance. It can be achieved by a broad diversity of commercially available and free softwares.[3–6] Protein identification, hence, is presented as routine pipelines due to development of good statistical algorithms.[7–9]

However, interpretation of the shotgun proteomics data is relatively considered as a challenge. Biological systems execute reactions and biological processes in functionality aspects, which rely on coordination of a bunch of proteins with related functions. Therefore, interpretation of large-scale data often includes looking for the biological functions that are enriched in lists of genes. Functional enrichment analysis uses statistical methods to discover significantly associated functional annotations, such as gene ontology terms, metabolic pathways, signaling pathways, protein-protein interactions, transcriptional regulations, post-transcriptional/post-translational modifications, and related diseases.[10–13]

In this chapter, we will demonstrate how to use two commonly used bioinformatics tools for functional enrichment analysis: (1) Gene Set Enrichment Analysis (GSEA)[12] and (2) the Database for Annotation, Visualization and Integrated Discovery (DAVID).[10] Moreover, we will use EnrichmentMap for network-based visualization of enriched functions.[14]

Preparation

— Download GSEA Java desktop application from the website: http://www.broadinstitute.org/gsea/index.jsp.
— Download and install Cytoscape from the website: http://www.cytoscape.org.
— Download and install the EnrichmentMap, a Cytoscape plugin, from the website: http://www.baderlab.org/Software/EnrichmentMap.

Data preparation for GSEA

To apply GSEA for enrichment analysis of proteome data, expression data and description of data are required.

— Expression data: TXT file (*.txt).

GSEA supports expression data in four formats: res, gct, pcl, or txt. For proteome data preparation, we recommend to create the expression data in tab delimited txt file format, which contains name (Gene Symbol), description, and expression value (the peak area of extracted ion current or $\log_2$ transformed value) in columns (Fig. 1). Convert the identifier of

	A	B	C	D	E	F
1	NAME	DESCRIPTION	CTL1	CTL2	MIR1	MIR2
2	AAAS	Q9NRG9	359518.917	366612.018	303987.924	300718.85
3	AACS	Q86V21	534105.745	562744.471	624819.462	661013.932
4	AAR2	Q9Y312	97935.5477	102784.05	116990.567	114028.722
5	AARS	P49588	14332404.2	14840283.4	10457297.3	10899981.3
6	AARSD1	Q9BTE6	1069156.32	972647.371	753046.049	744135.213
7	ABCD3	P28288	883326.051	881165.776	779161.17	794774.521
8	ABCE1	P61221	9193889.69	9311425.16	8656531.61	8920740.86
9	ABCF1	Q8NE71	4295055.26	4200988.8	4164511.64	4296864.79
10	ABCF2	Q9UG63	2354482.13	2176644.6	2247454.68	2236028.98
11	ABCF3	Q9NUQ8	797050.382	836847.807	750126.623	719952.251
12	ABHD12	Q8N2K0	303919.861	299440.706	256988.189	249751.939

Figure 1. The organization of the text file format for expression dataset. The first line contains the NAME and DESCRIPTION followed by the identifiers for each sample in the dataset. DESCRIPTION is optional; fill in it with "NA" if the description is not available. We use the Gene Symbol as the NAME, and the identified UniProt protein accession number as the DESCRIPTION for each expression value.

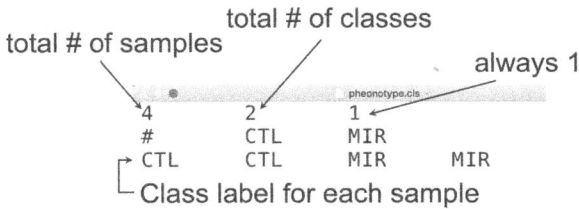

total # of classes

total # of samples

always 1

4	2	1
#	CTL	MIR
CTL	CTL	MIR	MIR

Class label for each sample

Figure 2. The organization of the cls file format for description of sample. The cls file consists of three lines, the number of samples and classes, the names of classes that have appeared in the analysis report, and the class label for each sample. Each label should be separated by a space or a tab in the cls file.

protein name into Official_Gene_Symbol using DAVID Gene ID Conversion Tool (http://david.abcc.ncifcrf.gov/conversion.jsp) if required. The data demonstrated in this chapter are iTRAQ labeled proteomics data which can be derived from the supplementary files from http://pubs.acs.org[15] or the ProteomeXchange Consortium via the PRIDE partner repository with the dataset identifier PXD001078.[16]

Description of data (phenotype): CLS file (*.cls)

The phenotype file assigns categorical class to each sample accordingly (e.g., control vs. treated, tumor vs. normal, or carcinoma *in situ* vs. metastasis, etc.). The file should be prepared in tab delimited txt file format (Fig. 2). The first line contains the total number of samples, the total number of classes, and 1. The second line contains a user-visible name for each class appearing in analysis reports. The line should begin with a pound sign (#) followed by a space or a tab. The third line contains a class label for each sample. The class label can be the class name, a number, or a text string. The order of label used here is same as the expression data and should be assigned as the same order of the category on the second line. For instance, CTL goes first and then MIR.

II. *Run GSEA*

Load expression and phenotype files (Fig. 3).

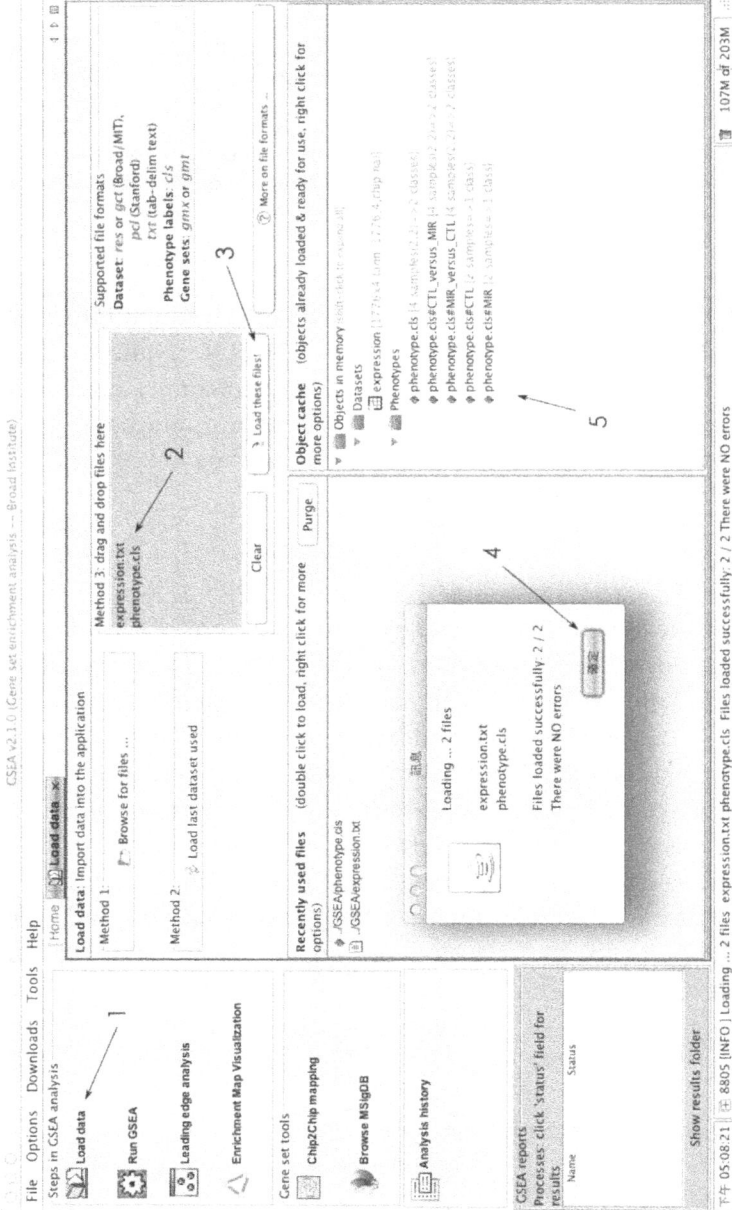

Figure 3. Load expression and phenotype files. 1. Click "Load data"; 2. Drag the required files into the GSEA application; 3. Click "Load these files"; 4. Click "Yes"; 5. The loaded files will be visible in the object cache.

Set parameters and Run (Fig. 4).

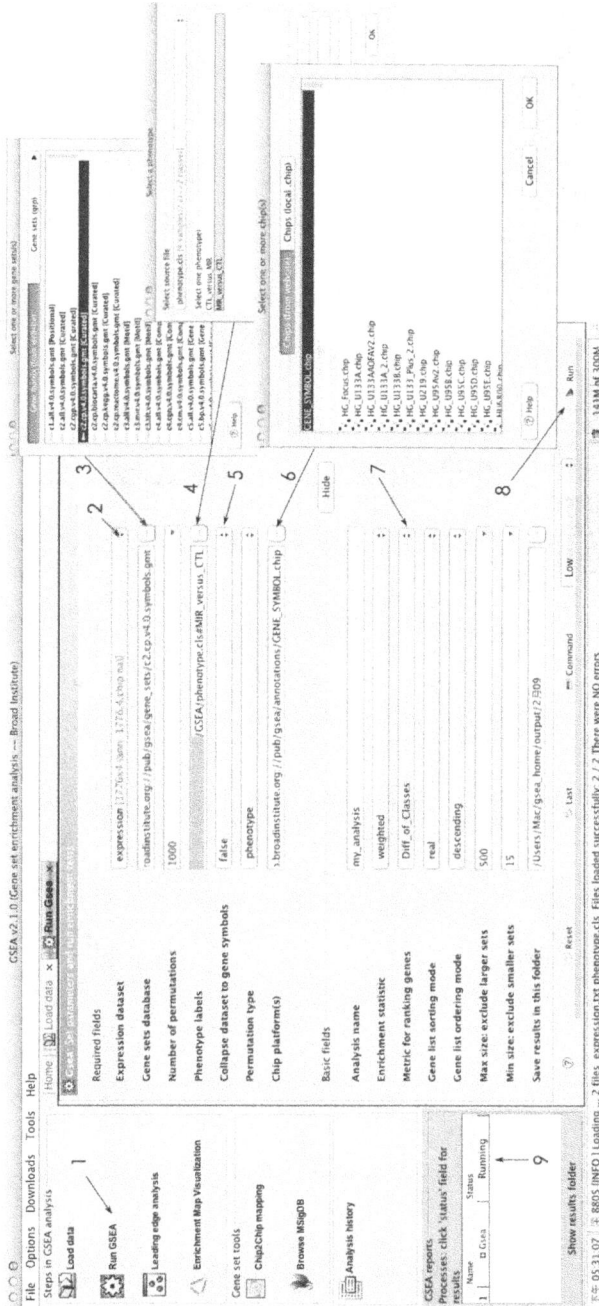

Figure 4. Set parameters applied for GSEA. 1. Click "Run GSEA"; 2. Select expression file; 3. Select the gene set of interests from the gene set database; 4. Select the phenotype labels for comparison; 5. Select "false" of collapse dataset to gene symbols; 6. Select "GENE_SYMBOL.chip" for annotation; 7. Select "Signal2Noise" for raw value or "Diff_of_Classes" for logarithm value to generate metric for ranking genes; 8. Click "Run"; 9. After the program is executed, the status of current jobs will be shown.

Example results of GSEA (Fig. 5).

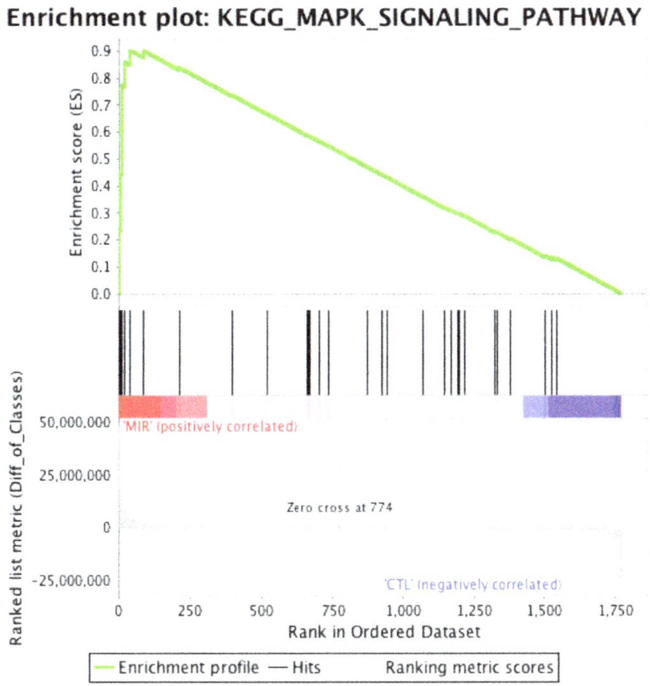

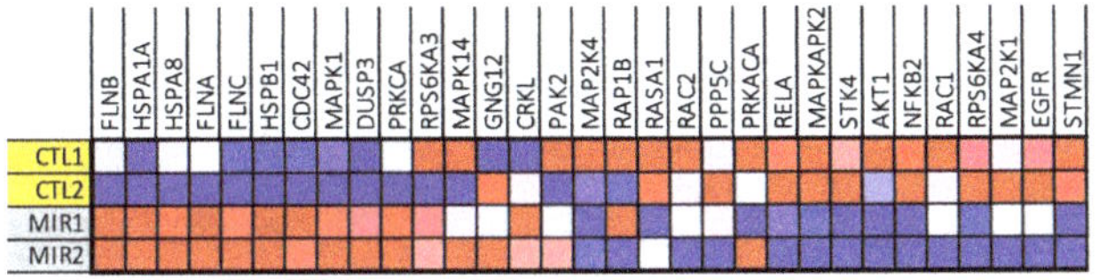

1	FLNB	O75369	FLNB Entrez, Source	filamin B, beta (actin binding protein 278)	3	20892368.000	0.2310	Yes
2	HSPA1A	P08107	HSPA1A Entrez, Source	heat shock 70kDa protein 1A	6	19227996.000	0.4441	Yes
3	HSPA8	P11142	HSPA8 Entrez, Source	heat shock 70kDa protein 8	10	15330410.000	0.6132	Yes
4	FLNA	P21333	FLNA Entrez, Source	filamin A, alpha (actin binding protein 280)	11	14344200.000	0.7730	Yes
5	FLNC	Q14315	FLNC Entrez, Source	filamin C, gamma (actin binding protein 280)	22	8084348.000	0.8574	Yes
6	HSPB1	P04792	HSPB1 Entrez, Source	heat shock 27kDa protein 1	41	4902909.000	0.9017	Yes
7	CDC42	P60953	CDC42 Entrez, Source	cell division cycle 42 (GTP binding protein, 25kDa)	88	2232963.750	0.9002	No
8	MAPK1	P28482	MAPK1 Entrez, Source	mitogen-activated protein kinase 1	212	799539.875	0.8385	No
9	DUSP3	P51452	DUSP3 Entrez, Source	dual specificity phosphatase 3 (vaccinia virus phosphatase VH1-related)	395	251329.813	0.7370	No
10	PRKCA	P17252	PRKCA Entrez, Source	protein kinase C, alpha	517	99229.188	0.6687	No
11	RPS6KA3	P51812	RPS6KA3 Entrez, Source	ribosomal protein S6 kinase, 90kDa, polypeptide 3	661	30219.125	0.5871	No
12	MAPK14	Q16539	MAPK14 Entrez, Source	mitogen-activated protein kinase 14	664	29101.938	0.5862	No

Figure 5. An example of GSEA result. Normalized enrichment score, enrichment plot, enriched gene list, and Blue-Pink O' Gram in the space of gene set can be obtained.

III. *Functional enrichment analysis using DAVID*

Upload the list of the differentially expressed proteins (Fig. 6).

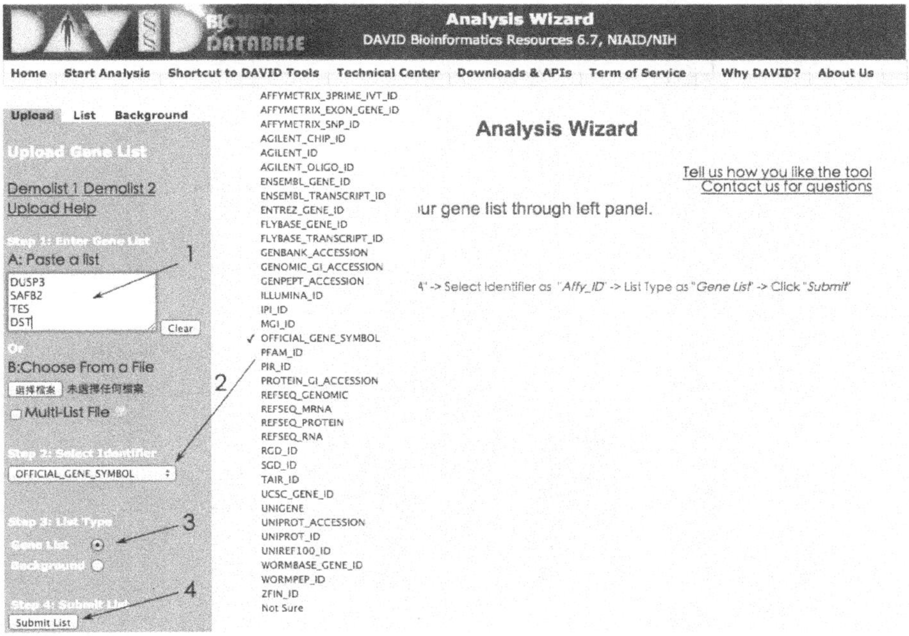

Figure 6. Upload the gene list to DAVID. Prepare the gene symbol of differentially expressed proteins in TXT file with a gene per line. 1. Copy/paste gene symbols to "box A"; 2. Select identifier as "OFFICIAL_GENE_SYMBOL"; 3. List type as "Gene List"; 4. Click "Submit" button.

Run and obtain the results of functional enrichment analysis (Fig. 7).

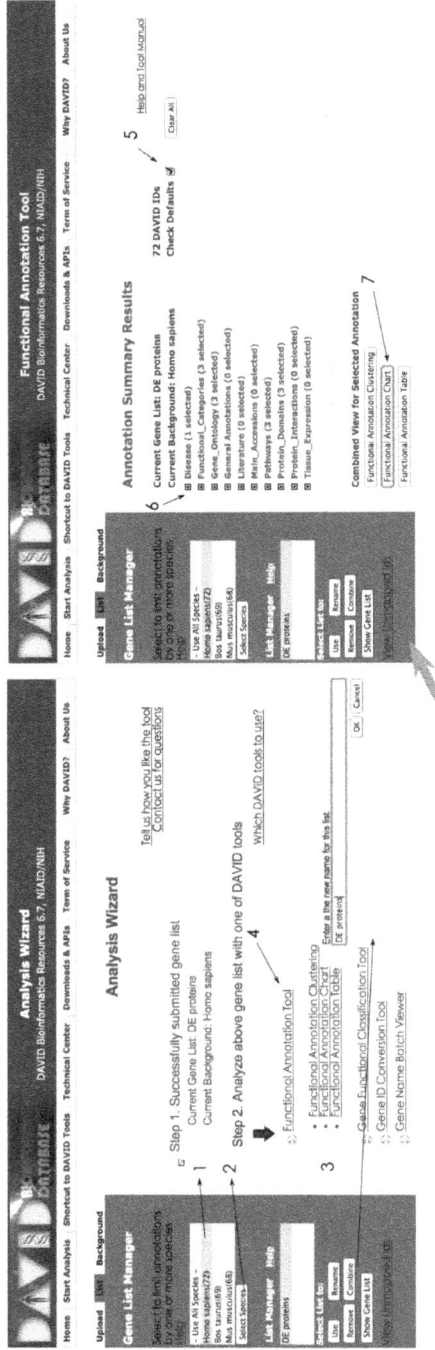

Figure 7. Select gene set for functional enrichment analysis. 1. Select species for the source of the input genes; 2. Click "Select Species"; 3. (Optional) Rename the list accordingly; 4. Click "Functional Annotation Tool"; 5. (Optional) Unselect "Check Defaults" if you prefer to perform functional enrichment analysis specifically; 6. (Optional) Select categories you want to enrich; 7. Click "Functional Annotation Charts" and save the results (Fig. 8).

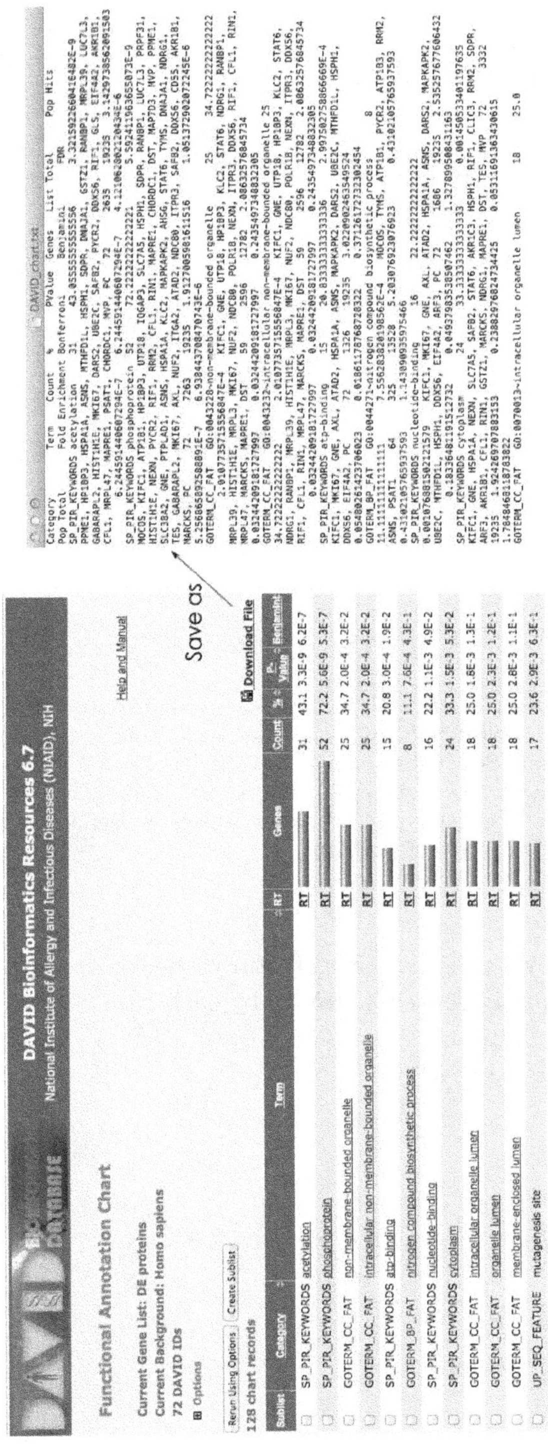

Figure 8. Save the results from DAVID. Save the chart results by right-clicking "Download File" and save the file.

IV. *Data interpretation using EnrichmentMap*

Interpret GSEA results (Fig. 9).

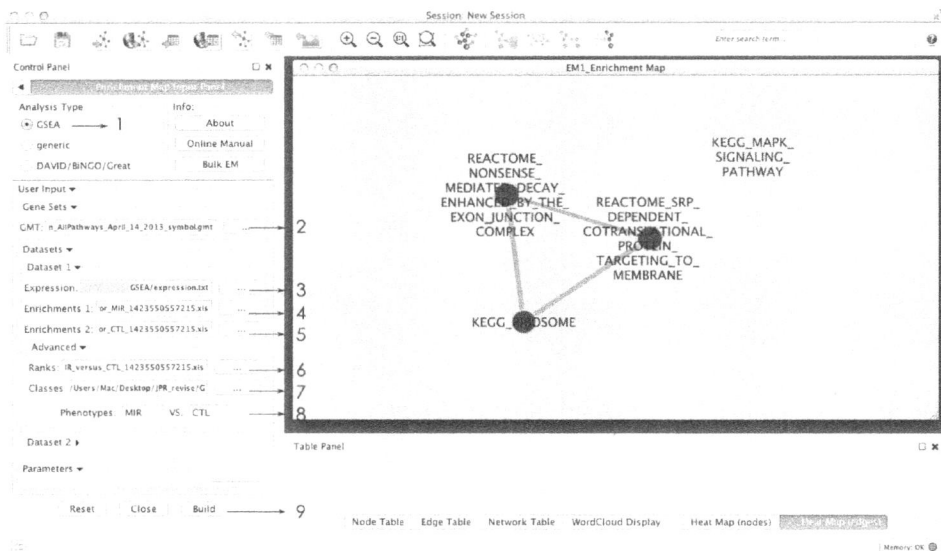

Figure 9. Interpret GSEA results. To perform EnrichmentMap of GSEA results, set the parameters as follows: 1. Select "GSEA" as analysis type; 2. Select the assigned GMT file used to perform GSEA; 3. Select the expression TXT file; 4 and 5. Select the enrichment results from GSEA output folder for each class; 6. Select the ranking file from GSEA output folder; 7. Select the phenotype CLS file as Classes; 8. Define the phenotype label; 9. Click "Build" to perform EnrichmentMap.

Launch Cytoscape program, install EnrichmentMap App from App > App Manager, and apply EnrichmentMap by click App > EnrichmentMap > Create Enrichment Map.

Interpret DAVID results (Fig. 10).

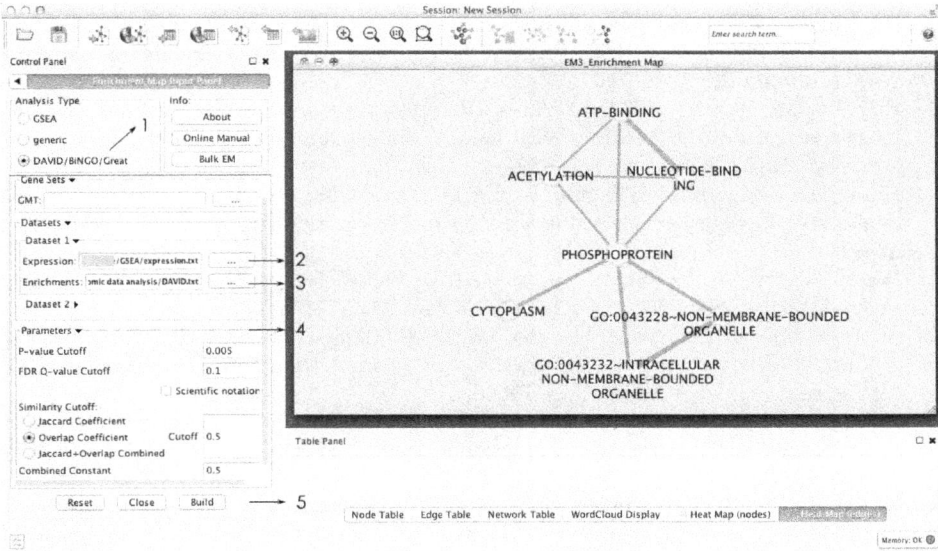

Figure 10. Interpret DAVID results. To perform EnrichmentMap of GSEA results, set the parameters as follows: 1. Select "DAVID/BiNGO/Great" as analysis type; 2. Select the expression TXT file; 3. Select DAVID result TXT file as the enrichments; 4. (Optional) Adjust the parameters for cut-off significant enriched functions and function similarity; 5. Click "Build" to perform EnrichmentMap.

Launch Cytoscape program, install EnrichmentMap App from App > App Manager, and apply EnrichmentMap by click App > EnrichmentMap > Create Enrichment Map.

References

1. Nilsson T, Mann M, Aebersold R *et al.* (2010) Mass spectrometry in high-throughput proteomics: Ready for the big time. *Nat. Methods* **7**: 681–685.
2. Mann M and Kelleher NL. (2008) Precision proteomics: The case for high resolution and high mass accuracy. *Proc. Natl. Acad. Sci. USA* **105**: 18132–18138.
3. Cox J and Mann M. (2008) Maxquant enables high peptide identification rates, individualized ppb-range mass accuracies and proteome-wide protein quantification. *Nat. Biotechnol.* **26**: 1367–1372.
4. Keller A, Eng J, Zhang N *et al.* (2005) A uniform proteomics ms/ms analysis platform utilizing open xml file formats. *Mol. Syst. Biol.* **1**: 2005.0017.
5. Rauch A, Bellew M, Eng J *et al.* (2006) Computational proteomics analysis system (cpas): An extensible, open-source analytic system for evaluating and publishing proteomic data and high throughput biological experiments. *J. Proteome Res.* **5**: 112–121.

6. Matthiesen R. (2007) In *Mass Spectrometry Data Analysis in Proteomics*, 121–138 (Springer).
7. Cottrell JS and London U. (1999) Probability-based protein identification by searching sequence databases using mass spectrometry data. *Electrophoresis* **20:** 3551–3567.
8. Craig R and Beavis RC. (2004) Tandem: Matching proteins with tandem mass spectra. *Bioinformatics* **20:** 1466–1467.
9. Eng JK, McCormack AL, and Yates JR. (1994) An approach to correlate tandem mass spectral data of peptides with amino acid sequences in a protein database. *J. Am. Soc. Mass. Spectro.* **5:** 976–989.
10. Huang DW, Sherman BT, and Lempicki RA. (2008) Systematic and integrative analysis of large gene lists using DAVID bioinformatics resources. *Nat. Protoc.* **4:** 44–57.
11. Wang J, Duncan D, Shi Z *et al.* (2013) Web-based gene set analysis toolkit (webgestalt): Update 2013. *Nucleic Acids Res.* **41:** W77–W83.
12. Subramanian A, Tamayo P, Mootha VK *et al.* (2005) Gene set enrichment analysis: A knowledge-based approach for interpreting genome-wide expression profiles. *Proc. Natl. Acad. Sci. USA* **102:** 15545–15550.
13. Huang DW, Sherman BT, and Lempicki RA. (2009) Bioinformatics enrichment tools: Paths toward the comprehensive functional analysis of large gene lists. *Nucleic Acids Res.* **37:** 1–13.
14. Merico D, Isserlin R, Stueker O *et al.* (2010) Enrichment map: A network-based method for gene-set enrichment visualization and interpretation. *PLoS One* **5:** e13984.
15. Chang HY, Li MH, Huang TC *et al.* (2015) Quantitative proteomics reveals middle infrared radiation-interfered networks in breast cancer cells. *J. Proteome Res.* **14:** 1250–1262.
16. Vizcaíno JA, Deutsch EW, Wang R *et al.* (2014) Proteomexchange provides globally coordinated proteomics data submission and dissemination. *Nat. Biotechnol.* **32:** 223–226.

7. Phosphorylation Data Analysis

Chia-Lang Hsu[*,‡] and Wei-Hsuan Wang[†]

*Department of Life Sciences,
National Taiwan University, Taipei, Taiwan
† Genome and Systems Biology Degree Program,
National Taiwan University, Taipei, Taiwan
‡ auymle@gmail.com

1. Introduction

Advances in mass spectrometry (MS) and enrichment methods allow large-scale measuring changes in site-specific protein phosphorylations at a high temporal resolution. A single MS-based phosphoproteomics experiment can generate datasets consisting of more than thousands of identified and quantified phosphorylation sites. The aim of performing such high-throughput phosphoproteomics investigation is to extract meaningful biological information that can provide mechanistic insights or hypotheses for further studies. Although this remains the bottleneck in the field of phosphoproteomics, the continuous developments and improvement of bioinformatics tools provide useful strategies to dissect large datasets and extract the biologically significant information. These tools can be used to determine, which kinase are more active under the experimental conditions,[1–3] which pathways or biological processes are significantly enriched in the data,[4,5] or to generate and visualize the data in the context of biological networks.[6–8]

To gain insight into the regulation of biological processes, it is required to monitor protein and phosphorylation dynamics by time-course or multiple-condition experiments. To unveil the dynamics behind the time-course or multiple-condition phosphoproteomics data, several bioinformatics tools have been developed to analyze such phosphoproteomics datasets.

PhosphoPath[6] is a Cytoscape app designed for the visualization of protein-protein interactions and site-specific kinase-substrate interactions with quantitative information for multiple conditions or time points. SELPHI[9] is a web-based server providing various functional analysis of phosphoproteomics data, such as enrichment analysis and correlation analysis.

In this chapter, we will demonstrate how to utilize DynaPho[10] to unveil the biological information behind the phosphoproteomics data. DynaPho is an integrative and web-based platform for analysis of large-scale phosphoproteomics data, especially time-course or multiple-condition experiments. DynaPho aims to provide the analysis of phosphoproteomics data readily available to the non-bioinformatics experts. Various functional analysis methods have been implemented in DynaPho, including phosphorylation profile clustering, temporal function enrichment analysis, kinase activity analysis, dynamic network analysis, and kinase/phosphatase-phosphosite association analysis. These methods have been comprehensively used in phosphoproteomics research as well as improved and extended for the investigation of time-series and multiple-condition data. In addition, DynaPho can record all analysis results with different parameter settings for a dataset. When a user uploads a phosphoproteomics dataset, a unique accession number is assigned to it and the user can retrieve their previous analysis results and resubmit an analysis with different parameter settings via this accession number. Through these different analysis results, users can interpret their data and generate a hypothesis for further studies.

Materials

Equipment requirements

DynaPho is accessed via the web interface and data analysis is performed on a dedicated web server. Thus, a standard computer with a modern web browser that supports JavaScript and HTML5 is required for the interaction of DynaPho interface.

Input data

After the processes of the peptide identification, quantification, and phosphosite estimation by other MS analysis softwares, such as MaxQuant,[11] users can extract the required information from the output files of these

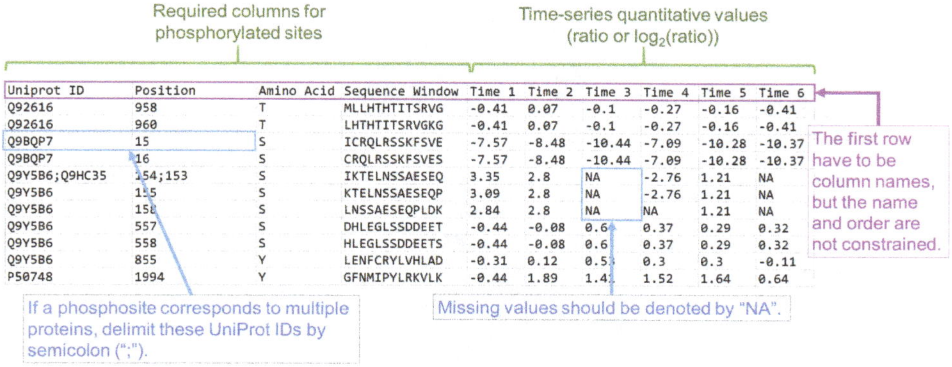

Figure 1. An example layout of quantitative phosphosite data for use in DynaPho.

software and prepare in the form of tab-delimited text files (Fig. 1). The required columns of input data for DynaPho include:

- UniProt ID. If a phosphorylated site maps to multiple proteins, the IDs are delimited by ";".
- Phosphorylation position on protein. For multiple hits, the values are delimited by ";".
- Phosphorylation residue (S, T, or Y).
- n-mer sequence window around the phosphorylated residues ($5 \leq n \leq 15$, where n is odd number). Empty positions for N- or C-terminal peptides have to be filled up manually with an underline character ("$-$") for each gap.
- Time-series quantitative values (Ratio). Base-2 logarithmic transformation of ratio value is recommended. The number of columns is based on your phosphoproteomics data.

Procedure

Submission of user's phosphoproteomic data

1. Start the web browser and load http://dynapho.jhlab.tw/.
2. Click on "New Job" on the header.
3. Choose the file and click on "Submit".
4. In "Edit Attributes" page, users can modify the column names and check whether the variable mapping is correct (Fig. 2). Once the settings are correct, click "Next" to start the data process.

(a)

(b)

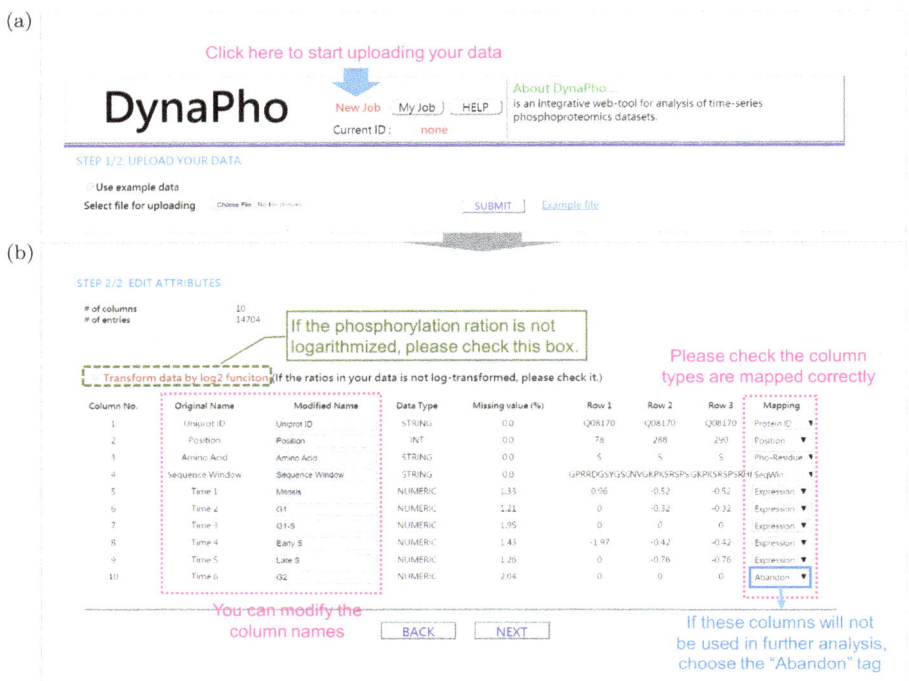

Figure 2. Submit data to DynaPho. The interface for (a) data selection and (b) data preview and attribution edition.

5. As the process is completed, the page will be directed to "Data Summary". In this page, DynaPho shows the distribution of phosphorylation residues by a pie chart and the distribution of phosphorylation ratios by a histogram. In addition, users can view their data and plot the profile of interested phosphosites (Fig. 3).

Clustering of phosphorylation profiles

The phosphosites with similar phosphorylation profiles might be involved in the same biological process or regulated by the same kinase family. To investigate the possible biological processes or regulators within the phosphoproteomics, the phosphosites will be grouped based on their phosphorylation profiles, and then the GO overrepresentation analysis and motif enrichment analysis will be performed on each group.

1. Click on "Profile Clustering" on the header.
2. Specify the method for missing value imputation (Fig. 4(a)). Because missing value is not allowed in clustering analysis, the phosphosites with

(a)

(b)

Distribution of phosphorylation measurements

(c)

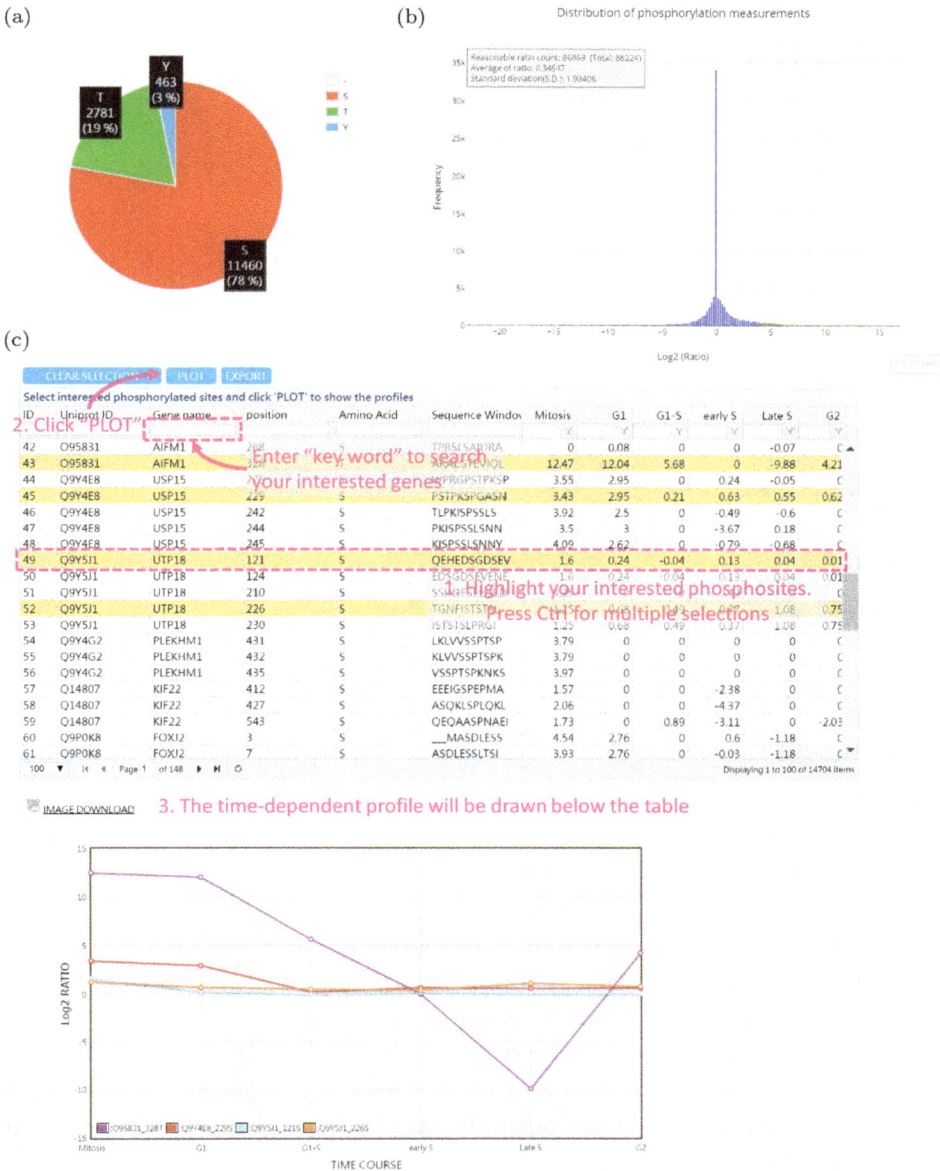

Figure 3. Data exploration of quantitative phosphosite data. DynapPho provides several ways for data exploration, including (a) the distribution of phosphorylation residues, (b) the distribution of phosphorylation ratios, and (c) the phosphorylation trend of a set of phosphosites.

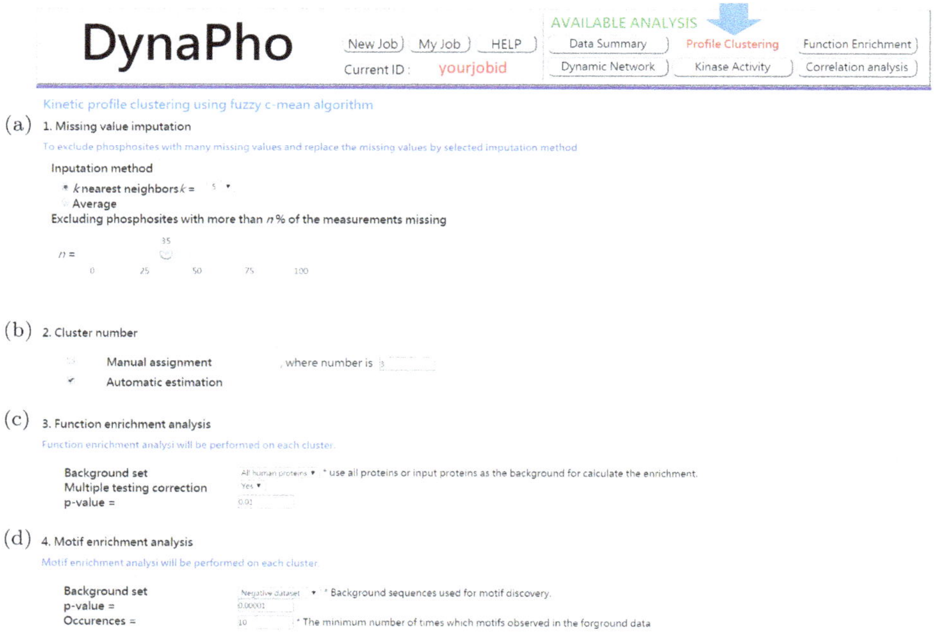

Figure 4.   The submission page of "profile clustering" analysis.

many missing values will be excluded and the remaining missing values will be imputed.

3. Specify the cluster number or select the automatic assignment. This is a required parameter for fuzzy c-mean algorithm.[12] Users can set a specific number or choose "automatic estimation" which will automatically assign an optimized number by the iterative procedures (Fig. 4(b)).

4. Specify the parameters for function enrichment analysis. Select a background set and define a p-value cutoff (Fig. 4(c)).

5. Specify the parameters for motif enrichment analysis. This analysis is performed by Motif-X algorithm.[13] Select a background set, set a minimum number of a motif occurring in the foreground data, and define a p-value cutoff (Fig. 4(d)).

6. Click on "Submit" to get to the result page.

7. Inspect profile clustering analysis results. The resulting page reveals each cluster by a profile plot (Fig. 5). Hit a specific profile plot to list the members of the cluster as well as the overrepresented motifs and GO terms. To visualize overrepresented GO terms, the GO terms and their overlap can be organized into an enrichment map. For more details about interpreting the enrichment map, please refer to the section "Construction of function profile".

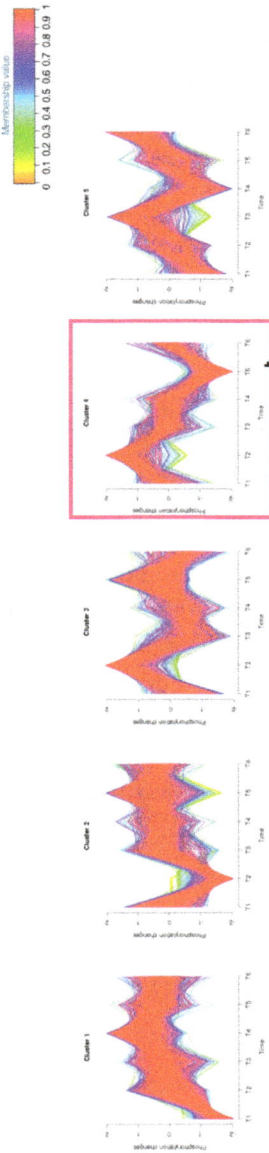

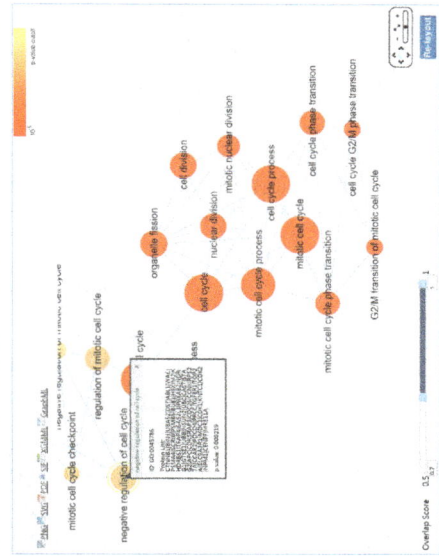

Figure 5. The result page of "profile clustering" showing clustered profiles and further functional analyses.

Construction of function profile

Gene set overrepresentation analysis is commonly used for the interpretation of proteomics and phosphoproteomics data. To reveal the dynamics of biological processes, DynaPho integrates the overrepresented gene set of all time point or conditions into a functional profile. In this analysis, DynaPho only considers the gene sets from biological process category of Gene Ontology (GO).

1. Click on "Function Enrichment" on the header.
2. Set a criterion for defining the proteins with differential change on phosphorylation.
3. Specify the parameters for function enrichment analysis. Select a background set and define a p-value cutoff.
4. Click on "Submit" to get to the result page.
5. Inspect function enrichment analysis results. For each time point or condition, the overrepresented GO terms are tabularized and graphed as an enrichment map, respectively (Fig. 6). The different GO terms and their overlap can be viewed in an interactive network. The node size is proportion to the number of members of the GO term, and the node color denotes enrichment significance. The GO terms with similar semantic meanings will tend to be connected and form a cluster. The complexity of the graph can be reduced by filtering edges based on the overlap score.
6. Inspect function profiles. DynaPho compiles the enrichment analysis results of all time points or conditions into a function profile and presents them as a heatmap. Hit the tab of "Function profile" to view the function profile (Fig. 7). The hierarchical cluster will be performed by enriched GO terms, and the GO terms with similar function profiles will be clustered together.

Construction of dynamic interaction networks

Protein–protein interaction network (PIN) can provide insights into signaling cascade. This analysis will generate PIN for each time point or condition, and user can compare the PINs among different time points or conditions to indicate the important interactions and pathways.

1. Click on "Dynamic Network" on the header.
2. Set a criterion for defining the proteins with differential change on phosphorylation. These proteins are used as seeds to construct the network.

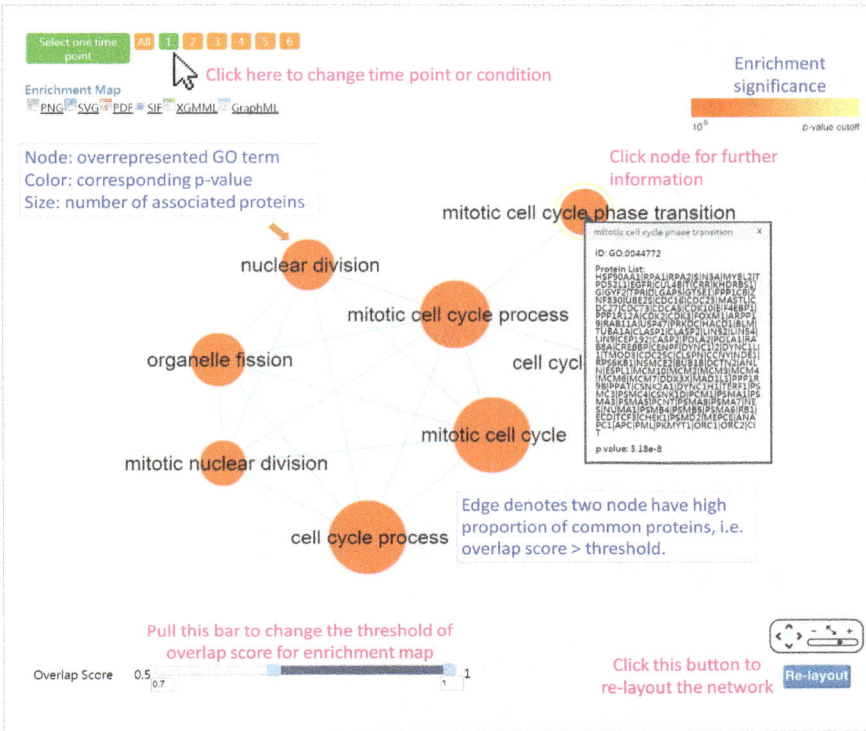

Figure 6. Visualization of overlap between enriched GO terms.

3. Specify a query condition. By default, DynaPho collects proteins which phosphorylation level are changed at least in one of time points. Users can focus on the proteins at a specific time point and uses these proteins as seeds to construct the network of each time point.

4. Click on "Submit" to get to the result page.

5. Inspect dynamic network analysis results. DynaPho provides two modes to view the networks: "proteins with phosphorylation sites" and "protein only" (Fig. 8). In "protein only" mode, the color of protein is based on the maximum phosphorylation change of the corresponding phosphosites. To view the transition of networks, DynaPho will fix the structure of the network, and users can change the time point or condition by dragging the slider bar (Fig. 9).

Kinase activity analysis

One of the major challenges in phosphoproteomics analysis is to identify the key kinases associated with the experimental conditions. The kinase

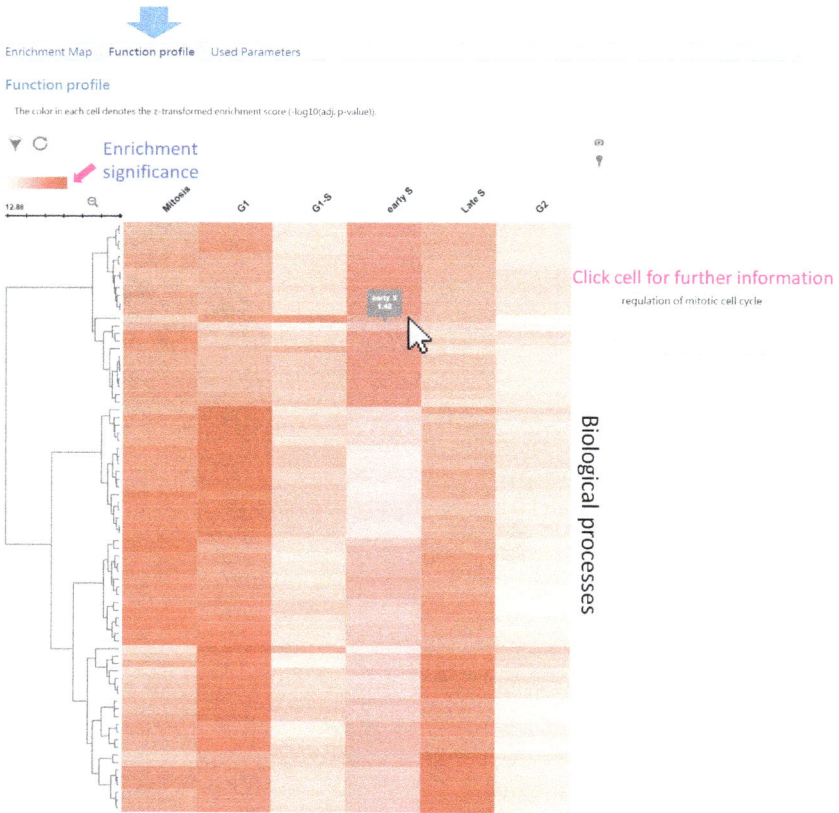

Figure 7. Visualization of function profile in DynaPho. The color of the cell indicates level of significance (darker = more significant).

activity profile is generated by integrating phosphosite residue with flanking sequences and phosphorylation profiles. This profile can reveal the activity of kinase at each time point or condition.

1. Click on "Kinase activity" on the header.
2. Set a criterion for defining the phosphosites with differential change on phosphorylation.
3. Set a *p*-value cutoff to determine the kinase with significant activity change.
4. Click on "Submit" to get to the result page.
5. Inspect kinase activity profile. The profile is visualized as a heatmap (Fig. 10). The kinases with similar profiles will be clustered. In the profile, the red and blue cells denote the kinases have significant

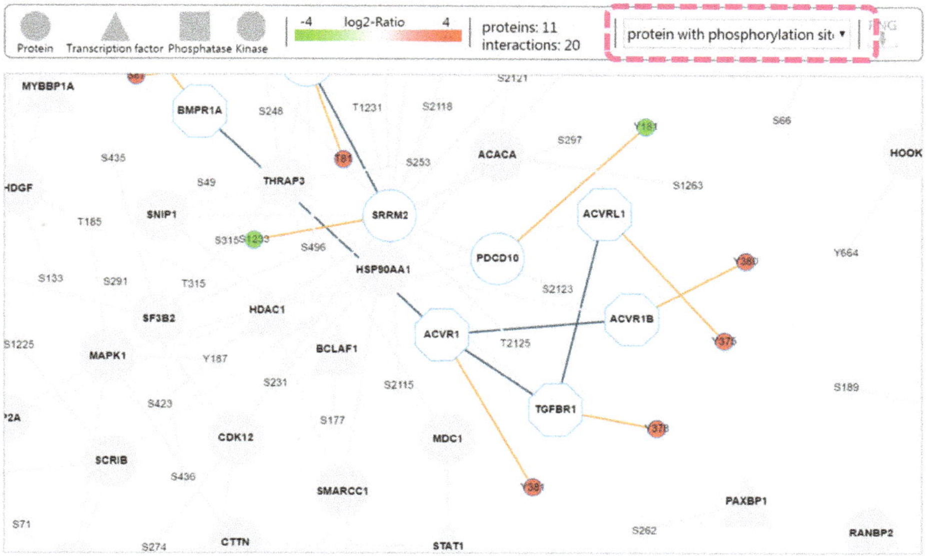

(a)

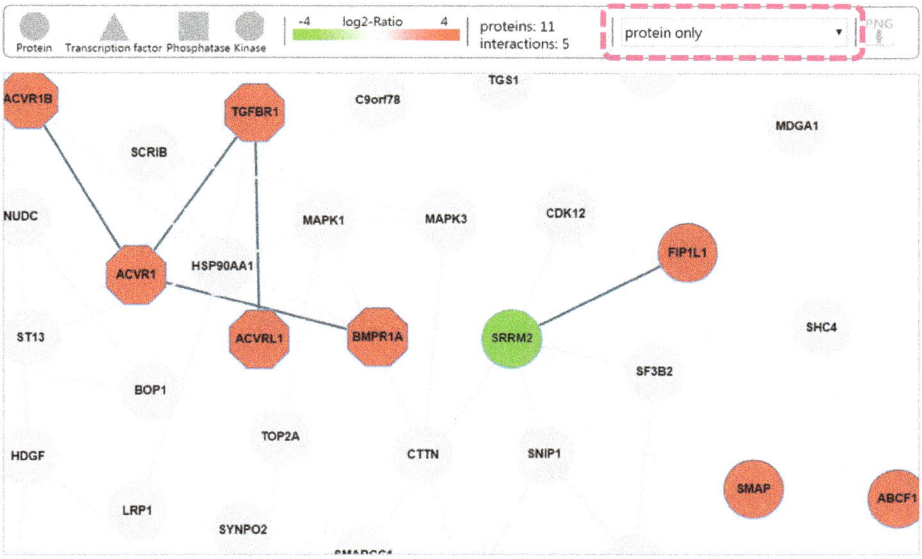

(b)

Figure 8. Interaction networks. DynaPho provides two kinds of network view: (a) showing protein and phosphorylation side and (b) showing protein only.

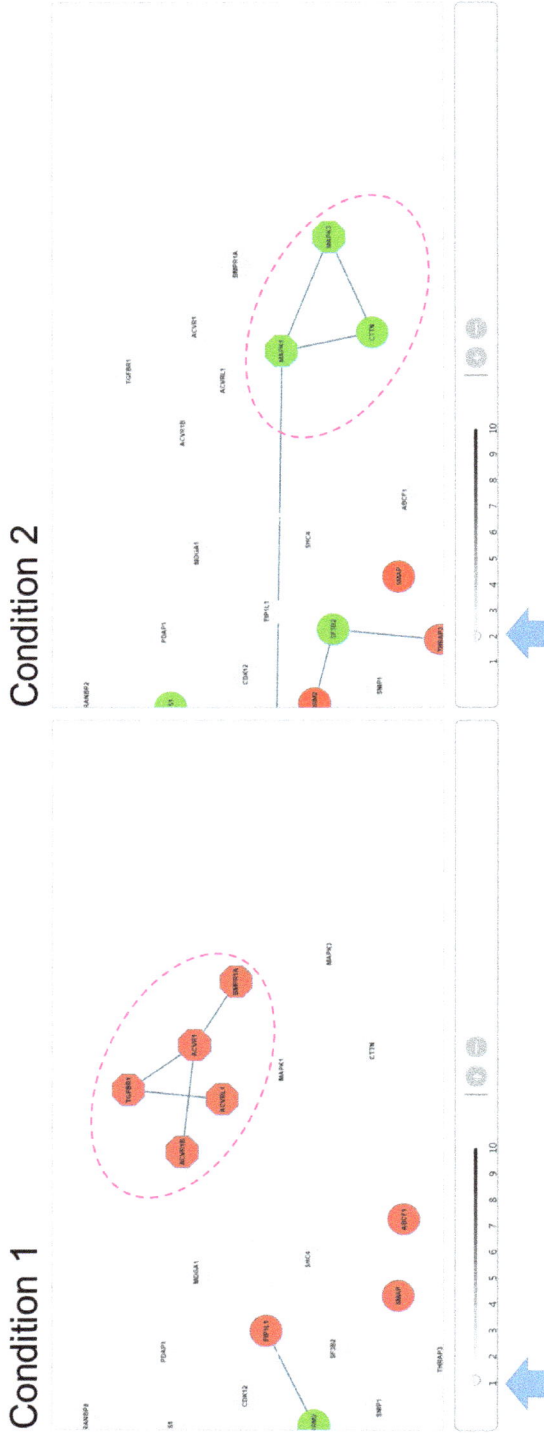

Figure 9.   Dynamic view of interaction networks. An example shows different conditions have different interaction modules.

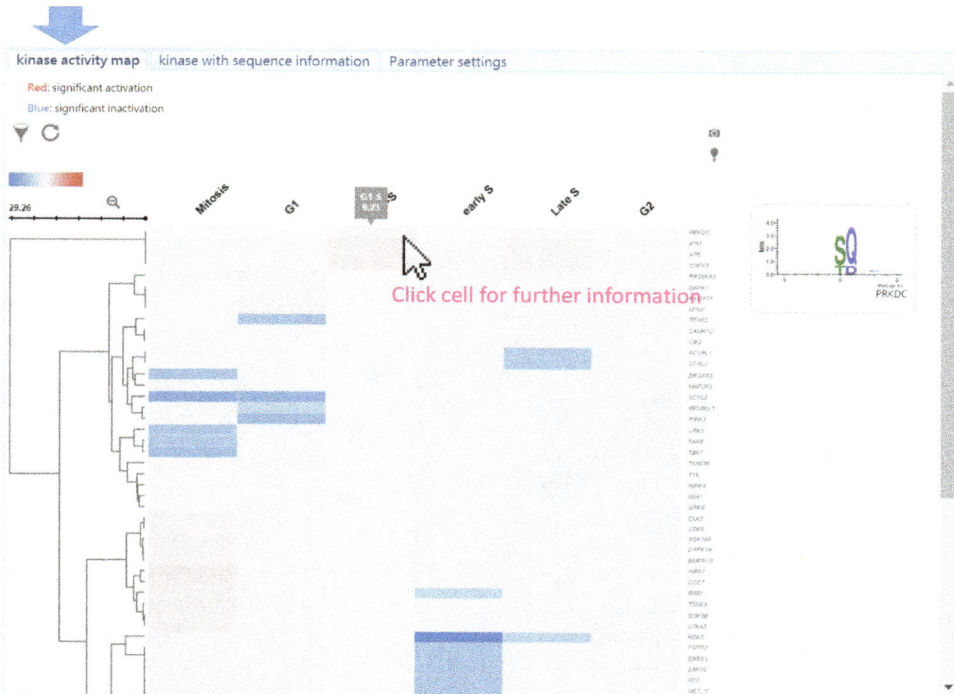

Figure 10. Visualization of kinase activity profile. The color of the cell indicates level of significance (darker = more significant) where blue and red are significant activation and inactivation and grey is no significance.

activation and inactivation at the given time point, respectively. The hierarchical clustering will be performed, and kinases with similar profiles will be clustered together. The consensus motif of a kinase will be show by dragging mouse on the row of this kinase.

Kinase/phosphatase–phosphosite association network

Kinase/phosphatase–substrate association is the basic component of the signaling transduction. To associate kinases or phosphatase with substrates identified in the form of individual phosphorylation sites, DynaPho applies a correlation-based method, which calculate the phosphorylation correlations to infer these associations.[9]

1. Click on "Correlation analysis" on the header.
2. Specify the parameters for correlation analysis. The kinase/phosphatase–phosphosite association network is constructed by a correlation-based method. To determine the relationship between kinase/phosphatase and

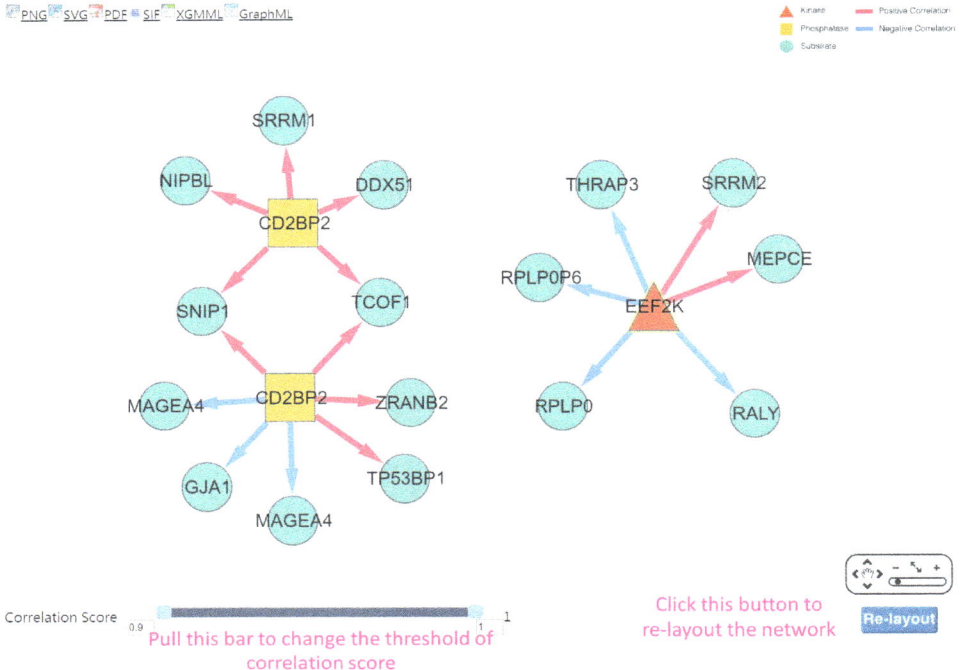

Figure 11.   Visualization of kinase/phosphatase–phosphosite association network.

phosphosite, the correlation value and the corresponding *p*-value are both considered. DynaPho requires a minimum number of three time points; however, users can set their data to only phosphosites that appear in a minimum number of time points via the parameter "Minimum samples", to be greater than required value. DynaPho provides two methods to calculate the correlation values: Spearman and Pearson correlation coefficients. The Spearman correlation is used by default.

3. Click on "Submit" to get to the result page.
4. Inspect kinase/phosphatase–phosphosite association network. The associations can be viewed in an interactive network (Fig. 11). The node color represents protein type, including kinase, phosphatase, and substrate. The edge color denotes the correlation type: positive (red) or negative (blue) correlation. The complexity of the graph can be reduce by filtering edges based on the correlation score.

Summary

DynaPho is a computational tool created with the aim of facilitating various analyses methods in a user-friendly and user-tailored way. In addition, differential analyses provide different information to assist users to easily interpret their phosphoproteomics data.

References

1. Locasale JW and Wolf-Yadlin A. (2009) Maximum entropy reconstructions of dynamic signaling networks from quantitative proteomics data. *PloS one.* **4**(8): e6522.
2. Ochoa D, Jonikas M, Lawrence RT *et al.* (2016) An atlas of human kinase regulation. *Mol. Syst. Biol.* **12**(12): 888.
3. Bennetzen MV, Cox J, Mann M *et al.* (2012) PhosphoSiteAnalyzer: A bioinformatic platform for deciphering phospho proteomes using kinase predictions retrieved from NetworKIN. *J. Proteome Res.* **11**(6): 3480–3486.
4. Huang da W, Sherman BT and Lempicki RA. (2009) Systematic and integrative analysis of large gene lists using DAVID bioinformatics resources. *Nat. Protoc.* **4**(1): 44–57.
5. Herwig R, Hardt C, Lienhard M *et al.* (2016) Analyzing and interpreting genome data at the network level with ConsensusPathDB. *Nat. Protoc.* **11**(10): 1889–1907.
6. Raaijmakers LM, Giansanti P, Possik PA, *et al.* (2015) PhosphoPath: Visualization of Phosphosite-centric Dynamics in Temporal Molecular Networks. *J. Proteome Res.* **14**(10): 4332–4341.
7. Linding R, Jensen LJ, Pasculescu A *et al.* (2008) NetworKIN: A resource for exploring cellular phosphorylation networks. *Nucleic Acids Res.* **36**(Database issue): D695–D699.
8. Szklarczyk D, Franceschini A, Wyder S *et al.* (2015) STRING v10: Protein-protein interaction networks, integrated over the tree of life. *Nucleic Acids Res.* **43**(Database issue): D447–D452.
9. Petsalaki E, Helbig AO, Gopal A *et al.* (2015) SELPHI: Correlation-based identification of kinase-associated networks from global phospho-proteomics data sets. *Nucleic Acids Res.* **43**(W1): W276–W282.
10. Hsu CL, Wang JK, Lu PC *et al.* (2017) DynaPho: A web platform for inferring the dynamics of time-series phosphoproteomics. *Bioinformatics.* doi:10.1093/bioinformatics/btx443.
11. Cox J and Mann M. (2008) MaxQuant enables high peptide identification rates, individualized p.p.b.-range mass accuracies and proteome-wide protein quantification. *Nat. Biotechnol.* **26**(12): 1367–1372.
12. Kumar L and Futschik ME. (2007) Mfuzz: A software package for soft clustering of microarray data. *Bioinformation.* **2**(1): 5–7.
13. Schwartz D and Gygi SP. (2005) An iterative statistical approach to the identification of protein phosphorylation motifs from large-scale data sets. *Nat. Biotechnol.* **23**(11): 1391–1398.

8. Pathway and Network Analysis

Chen-Tsung Huang and Hsueh-Fen Juan*

*Graduate Institute of Biomedical Electronics and Bioinformatics,
National Taiwan University, Taiwan
yukijuan@ntu.edu.tw

Pathway and network analysis allows us to integrate diverse 'omic' data (genomics, transcriptomics, proteomics, etc.) and interpret them into more comprehensible biological pathways and networks that possibly underlie a given complex biological system (e.g., certain cancers). In this chapter, we illustrate the pathway and network analyses using the R programming language, the most popular object-oriented statistical programming language among biostatisticians and biomedical faculty with numerously available R packages contributed by users around the world. These step-by-step procedures would help you get more familiar with the skills and techniques in cancer systems biology.

1. Installation of R

You can go to https://www.r-project.org/ to download the latest version of R. After R is properly installed, you can get started by opening an R session and entering the commands as you want.

2.   Installation of relevant R packages

Several R packages have been made available via Bioconductor,[1] an open source framework for bioinformatics implemented in R. To source Bioconductor in the R session, enter the following command:

> *source("http://www.bioconductor.org/biocLite.R")*

Then you can install specific packages, e.g., an R package ALL, which contains a gene expression dataset about patients with leukemia,[2] via the function biocLite() as follows:

> *biocLite("ALL")*

The package ALL will be downloaded into your computer. Other R packages that we use in this chapter include hgu95av.db, org.Hs.eg.db, samr, GOstats, KEGG.db, and Rgraphviz, and should be downloaded with the following commands:

> *biocLite("hgu95av.db")*
> *biocLite("org.Hs.eg.db")*
> *biocLite("samr")*
> *biocLite("GOstats")*
> *biocLite("KEGG.db")*
> *biocLite("Rgraphviz")*

3.   Getting the gene expression dataset

To get started with a gene expression dataset, load the package ALL using the function library():

> *library(ALL)*

Now, ALL package has been loaded to your computer, but not to R. To see all available data sets in the ALL package, enter the command:

> *data(package="ALL")*

You can see there is a data set called ALL in this package. To load this data set into R, use the command:

> *data(ALL)*

Enter the command to learn more details about this data set ALL:

> *ALL*

ExpressionSet (storageMode: lockedEnvironment)
assayData: 12625 features, 128 samples

element names: exprs

protocolData: none

phenoData

      sampleNames: 01005 01010 ... LAL4 (128 total)

      varLabels: cod diagnosis ... date last seen (21 total)

      varMetadata: labelDescription

featureData: none

experimentData: use 'experimentData(object)'

      pubMedIds: 14684422 16243790

Annotation: hgu95av2

Now, you know that ALL is an object of ExpressionSet and annotated under the microarray platform hgu95av. To fetch the expression matrix in ALL of the class ExpressionSet, use the function exprs():

>*DATA.ALL <- exprs(ALL)*

Next, you can assign a matrix named DATA.ALL from the object ALL, where each row corresponds to a probe name and each column represents a patient (sample).

4. Probe-to-gene conversion of the gene expression dataset

To facilitate the subsequent analyses, you can transform the probe sets of any microarray platform into the corresponding gene symbols if a mapping is available. To this end, considering the fact that ALL is annotated under the microarray platform hgu95av, you can load the R package hgu95av.db which has been downloaded beforehand:

>*library(hgu95av.db)*

You can check all available objects in hgu95av.db using the command:

>*ls("package:hgu95av2.db")*

You can use two Bimap objects hgu95av2ENTREZID and hgu95av2SYM BOL in this package, which store the mappings between probe labels and Entrez gene identifiers[3] and between probe labels and gene symbols, respectively, to get the mapped Entrez IDs and gene symbols (NA is assigned if there's no mapping for that probe):

>*ez <- mget(row.names(DATA.ALL), envir = hgu95av2ENTREZID)*
>*ez <- as.character(sapply(ez, "["))*
>*sb <- mget(row.names(DATA.ALL), envir = hgu95av2SYMBOL)*

>sb <- as.character(sapply(sb, "["))
>names(sb) <- ez

Note that any time if you feel confused about the usage of a function, say mget, type help(mget) to learn more details about the function mget. Next, you could get a vector of strings that combine each gene symbol and corresponding Entrez ID separated by a vertical bar "|":

>sbez <- paste(sb, ez, sep= "/")

Now, you can replace the old row names in the DATA.ALL matrix with the new gene names sbez and order the rows according to the new row names alphabetically:

>rownames(DATA.ALL) <- sbez
>DATA.ALL <- DATA.ALL[!is.na(ez),]
>DATA.ALL <- DATA.ALL[order(row.names(DATA.ALL)),]

5. Taking median values for multiple genes of the transformed dataset

After DATA.ALL is transformed into a gene-based matrix, you can next handle the problem of multiple expression values corresponding to the same genes. A solution to this problem is usually taking the median expression values for the same genes in each sample (patient) on the log2-tranformed basis. Now, you can define a function median4MultiRows that takes an input matrix with row names and returns an output matrix, where row names are ordered and unique, that takes into account median values for multiple row names of the input matrix:

>median4MultiRows <- function(mtrx) {
+   uniqueNames <- unique(rownames(mtrx))
+   uniqueNames <- uniqueNames[order(uniqueNames)]
+   out <- NULL
+   for (nm in uniqueNames)
+   out <- rbind(out, apply(mtrx[rownames(mtrx) %in% nm,, drop=F],
       2, median))
+   rownames(out) <- uniqueNames
+   return(out)
+}

Now, you can use this function to resolve the multiple gene name problem:

>DATA.ALL <- median4MultiRows(DATA.ALL)

6. Sample selection

At this stage, you are getting interested in the tumor biology of acute lymphoblastic leukemia (ALL) with different kinds of molecular genetic abnormalities. There were several metadata delineating additional clinical information in the ExpressionSet object ALL. ALL$BT shows you the types of leukemic lymphocytes (B or T cells) with the stages of diseases (1–4). ALL$mol indicates whether these leukemic samples harbored some common genetic abnormalities, such as BCR−ABL[4,5] (representative of the Philadelphia chromosome[6]) or ALL1−AF4[7] fusion genes. Here, you want to compare B cell leukemia with BCR−ABL rearrangement and B cell leukemia with ALL1−AF4 rearrangement, and fetch the relevant sample indices as follows:

```
>bCell <- grep("^B", as.character(ALL$BT))
>bcrAbl <- which(ALL$mol == "BCR/ABL")
>all1Af4 <- which(ALL$mol == "ALL1/AF4")
```

You can then take intersections between bCell and bcrAbl, and between bCell and all1Af4 as well as select the corresponding gene expression matrices from DATA.ALL, respectively:

```
>data.bcrabl <- DATA.ALL[, intersect(bCell, bcrAbl)]
>data.all1Af4 <- DATA.ALL[, intersect(bCell, all1Af4)]
```

You can get the number of samples and rename these samples as follows:

```
>N.bcrabl <- ncol(data.bcrabl)
>N.all1Af4 <- ncol(data.all1Af4)
>colnames(data.bcrabl) <- paste("BCR/ABL", 1:N.bcrabl , sep= "_")
>colnames(data.all1Af4) <- paste("ALL1/AF4", 1:N.all1Af4, sep= "_")
```

7. Analysis of differentially expressed genes

In this step, we apply a method called Significant Analysis of Microarray (SAM)[8] to B cell leukemia with BCR/ABL or ALL1/AF4 rearrangement and find out the differentially expressed (DE) genes between them. Here, you can use the R package samr, which is able to implement several functions necessary for SAM. Before taking advantage of the function samr(), you can prepare the input format for this function:

```
>library(samr)
>mergedData <- cbind(data.bcrabl, data.all1Af4)
>classLabel <- c(rep(1, N.bcrabl), rep(2, N.all1Af4))
```

>*samrData <- list(x= mergedData,*
+   *y = classLabel,*
+   *genenames= rownames(mergedData),*
+   *logged2= T)*

where you designate BCR/ABL samples as class label 1 and ALL1/AF4 samples as class label 2. Now, you can perform SAM using the two class unpaired standard t-test with 1000 permutations:

>*samrObj <- samr(samrData,*
+   *resp.type = "Two class unpaired",*
+   *testStatistic = "standard",*
+   *nperms =1000,*
+   *assay.type = "array")*

Note that you can do other adjustments to the input arguments for your SAM test using help(samr) for more details. The returned object is a list of several values and stored as samrObj. Next, you can use the function samr.compute.delta.table() to compute the delta values and false discovery rates (FDRs):

>*samrDeltaTable <- samr.compute.delta.table(samrObj)*

You can then select a minimum delta value with median FDR <0.01 according to samrDeltaTable:

>*myDelta <- min(samrDeltaTable[which(samrDeltaTable[, "median FDR"] <0.01), "delta"])*

Now, you can draw a Q–Q plot according to this delta value, showing only DE genes with fold change greater than or equal to two (red point for DE upregulation when class label 1 is used as the reference group; green for downregulation, Fig. 1):

>*samr.plot(samrObj, del=myDelta, min.foldchange=2)*

Then you can compute a table of significant genes using the function samr.compute.siggenes.table() taking: samrObj, myDelta, samrData, and samrDeltaTable as arguments:

>*samrSummTable <- samr.compute.siggenes.table(samrObj, del=myDelta, samrData, samrDeltaTable)*

You can show the number of DE upregulated genes in class label 2 (i.e., ALL1/AF4 samples) compared to class label 1 (i.e., BCR/ABL samples):

>*samrSummTable$ngenes.up*
[1] 156

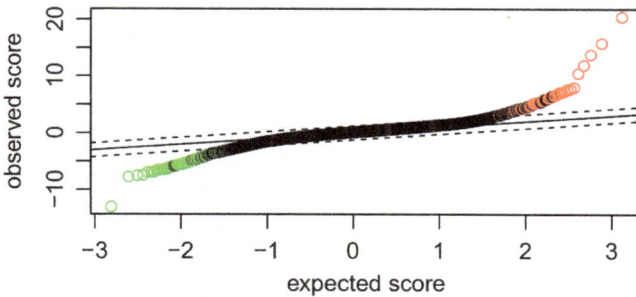

Figure 1. QQ plot for SAM analysis. Red points represent DE upregulated genes (when class label 1 is use as the reference group) while green DE downregulated genes.

Ditto for DE downregulated genes:

>samrSummTable$ngenes.lo
[1] 309

Note that the number of DE genes *might* be different each time when you perform a SAM test due to the nature of the permutation procedure. We can show the first parts of the table for DE genes using head():

>head(samrSummTable$genes.up)
>head(samrSummTable$genes.lo)

From the output in the screen, you learn that "CCNA|8900" is the most significantly upregulated gene in with fold change = 8.43 and "CD52|1043" is the most significantly downregulated gene with fold change = 0.08 in ALL1/AF4 samples (class label 2) compared to BCR/ABL samples (class label 1). You can go back to check the input expression matrix for these two genes:

>mergedData["CCNA1/8900",]
>mergedData["CD52/1043",]

8. KEGG pathway analysis

After you obtain DE genes between BCR/ABL and ALL1/AF4 samples using SAM analysis, you may want to ask whether these DE genes can be enriched in some molecular pathways that discern these two types of B cell leukemia. Here, you can perform a hypergeometric test on Kyoto Encyclopedia of Genes and Genomes (KEGG)[9] pathways using Entrez identifiers of the DE genes. Before that, you have to obtain a mapping between Entrez identifiers and KEGG identifiers using the Bimap object org.Hs.egPATH in R package org.Hs.eg.db:

```
>library(org.Hs.eg.db)
>x <- org.Hs.egPATH
>mappedIDToKEGG <- mappedLkeys(x)
>Gos <- mget(mappedIDToKEGG, org.Hs.egPATH, ifnotfound =NA)
>haveKEGG <- as.vector(!sapply(Gos, function(x) any(is.na(x)),
   simplify= T))
>mappedIDToKEGG <- mappedIDToKEGG[haveKEGG]
```

mappedIDToKEGG is a vector of Entrez IDs that have at least one mapping
to the KEGG pathways. Next, you may want to fetch Entrez IDs from the
previous SAM results stored in samrSummTable:

```
>up <- samrSummTable$genes.up[,2]
>up <- sapply(strsplit(up, "\\/"), "[")[2,]
>dn <- samrSummTable$genes.lo[,2]
>dn <- sapply(strsplit(dn, "\\/"), "[")[2,]
```

Now, you also have to derive a gene universe for the hypergeometric test.
A typical selection is the whole Entrez IDs used in the SAM test:

```
>universe <- unique(ez)
```

You can then restrict these Entrez IDs to those having at least one KEGG
mapping:

```
>up <- up[up %in% mappedIDToKEGG]
>dn <- dn[dn %in% mappedIDToKEGG]
>universe <- universe[universe %in% mappedIDToKEGG]
```

After Entrez IDs get prepared, you can start the hypergeometric test
using the function hyperGTest() in R package GOstats.[10] The input for
hyperGTest() takes a KEGGHyperGParams object with several arguments
as follows:

```
>library(GOstats)
>params <- new("KEGGHyperGParams",
+              geneIds = up,
+              universeGeneIds = universe,
+              annotation = "org.Hs.eg.db",
+              pvalueCutoff = 0.05,
+              testDirection = "over")
```

where genelds takes the Entrez IDs to be tested, universeGeneIds takes the
universe of Entrez IDs, annotation indicates the R package for annotation,
pvalueCutoff is set to 0.05, and testDirection determines overrepresentation
or underrepresentation of the test. This KEGGHyperGParams object is

then subjected to the hypergeometric test on overrepresentation in KEGG pathways for those upregulated genes in ALL1/AF4 samples (or called downregulated genes in BCR/ABL samples):

```
>hgOver <- hyperGTest(params)
>df.up <- summary(hgOver)
>head(df.up)
```

You can also test the downregulated genes in ALL1/AF4 samples (or called upregulated genes in BCR/ABL samples) under the same settings. You can create another KEGGHyperGParams object or simply adjust the parameter geneIds in the KEGGHyperGParams object params and test as follows:

```
>geneIds(params) <- dn
>hgOver <- hyperGTest(params)
>df.dn <- summary(hgOver)
>head(df.dn)
```

Or you can test the union of both upregulated and downregulated genes like this:

```
>geneIds(params) <- c(up, dn)
>hgOver <- hyperGTest(params)
>df.both <- summary(hgOver)
>head(df.both)
```

From these pathway analyses, you can observe that KEGG terms like "Cell adhesion molecules (CAMs)" or "Antigen processing and presentation" are enriched for downregulated genes in ALL1/AF4 samples (or upregulated genes in BCR/ABL samples), suggesting that BCR/ABL-rearranged B cell leukemia may be more immunoreactive than ALL1/AF4-rearranged B cell leukemia.

9. Creating gene correlation networks

To create gene–gene correlation networks for these BCR/ABL and ALL1/AF4 leukemia samples, you can use Pearson correlation coefficients (PCCs) and DE genes derived from the SAM analysis. You can utilize the function compCorrGraph() in R package GOstats to compute PCCs of genes. Additionally, you can specifically take the DE genes with fold change greater than or equal to 4 into consideration. In order to be able to plot the graph object returned by compCorrGraph(), you have to load the package Rgraphviz:

```
>library(Rgraphviz)
```

$>up.4.fold$ <- $samrSummTable\$genes.up[,2][as.numeric(samrSummTable$
$\$genes.up[,7]) >=4]$
$>dn.4.fold$ <- $samrSummTable\$genes.lo[,2][as.numeric(samrSummTable$
$\$genes.lo[,7]) <=1/4]$

The compCorrGraph() takes an ExpressionSet object as the argument and the parameter tau is the cutoff beyond which absolute PCC values will be plotted in the graph as visible undirected edges. Here, you can plot a correlation graph in ALL1/AF4 samples (N=10) for these DE genes under |PCC| >0.7 (corresponding to a two-tailed P-value = 0.024; Fig. 2):

$>eMat$ <- $data.all1Af4[rownames(data.all1Af4)$ %in% $c(up.4.fold, dn.4.$
    $fold),]$
$>rownames(eMat)$ <- $sapply(strsplit(rownames(eMat),$ "\\/"), "[")[1,]$
$>eSet$ <- $new("ExpressionSet", exprs=eMat)$
$>corrG$ <- $compCorrGraph(eSet, tau=0.7)$
$>edgemode(corrG)$ <- $"undirected"$
$>plot(corrG)$

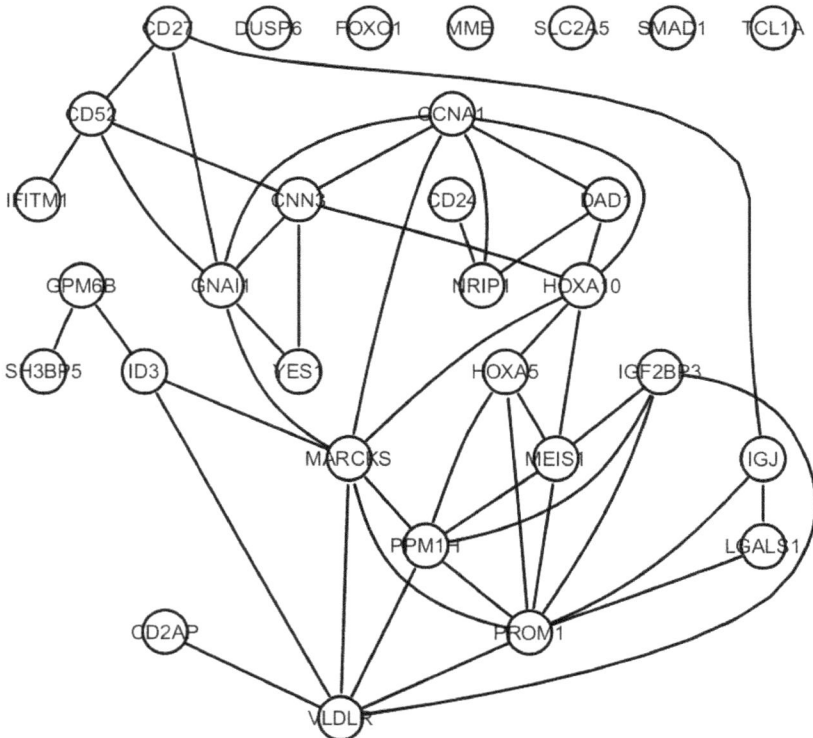

Figure 2.   Gene correlation network for B cell leukemia with ALL1−AF4 rearrangement.

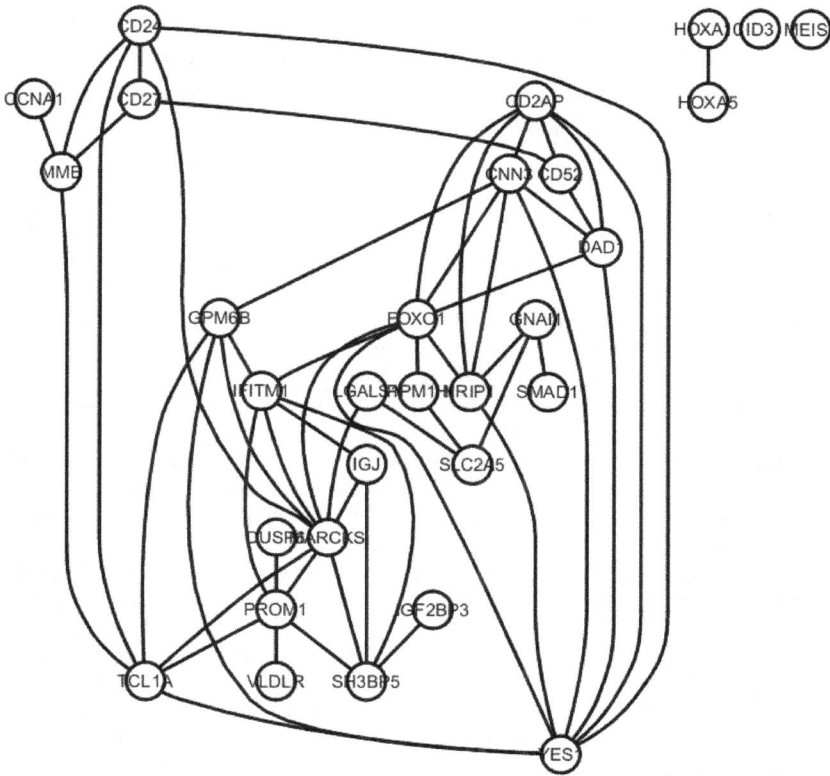

Figure 3.   Gene correlation network for B cell leukemia with BCR−ABL rearrangement.·

Also, you can plot a correlation graph in BCR/ABL samples (N=37) for these DE genes under |PCC|>0.4 (corresponding to a two-tailed P-value = 0.014; see Fig. 3):

```
>eMat <- data.bcrabl[rownames(data.bcrabl) %in% c(up.4.fold, dn.4.fold),]
>rownames(eMat) <- sapply(strsplit(rownames(eMat), "\\/"), "[")[1,]
>eSet <- new("ExpressionSet", exprs=eMat)
>corrG <- compCorrGraph(eSet, tau=0.4)
>edgemode(corrG) <- "undirected"
>plot(corrG)
```

10.   Summary and perspective

In this chapter, you have learned some techniques about pathway and network analysis to clearly see the genomic differences between B cell leukemia with ALL1−AF4 and BCR−ABL rearrangements. With more advanced approaches, you will get closer to a comprehensive picture about the complex biological systems in individual cancers that can ultimately translate into improvements in the treatment and general well-being for patients who suffer from these diseases.

References

1. Reimers M and Carey VJ. (2006) Bioconductor: An open source framework for bioinformatics and computational biology. *Method Enzymol.* **411:** 119–134.
2. Chiaretti S, Li X, Gentleman R *et al.* (2004) Gene expression profile of adult T-cell acute lymphocytic leukemia identifies distinct subsets of patients with different response to therapy and survival. *Blood* **103:** 2771–2778.
3. Maglott D, Ostell J, Pruitt KD *et al.* (2011) Entrez Gene: Gene-centered information at NCBI. *Nucleic Acids Res.* **39:** D52–57.
4. Gleissner B, Gokbuget N, Bartram CR *et al.* (2002) Leading prognostic relevance of the BCR-ABL translocation in adult acute B-lineage lymphoblastic leukemia: A prospective study of the German Multicenter Trial Group and confirmed polymerase chain reaction analysis. *Blood* **99:** 1536–1543.
5. Radich JP, Kopecky KJ, Boldt DH *et al.* (1994) Detection of BCR-ABL fusion genes in adult acute lymphoblastic leukemia by the polymerase chain reaction. *Leukemia* **8:** 1688–1695.
6. Lozzio CB and Lozzio BB. (1975) Human chronic myelogenous leukemia cell-line with positive Philadelphia chromosome. *Blood* **45:** 321–334.
7. Biondi A, Rambaldi A, Rossi V *et al.* (1993) Detection of ALL-1/AF4 fusion transcript by reverse transcription-polymerase chain reaction for diagnosis and monitoring of acute leukemias with the t(4;11) translocation. *Blood* **82:** 2943–2947.
8. Tusher VG, Tibshirani R, and Chu G. (2001) Significance analysis of microarrays applied to the ionizing radiation response. *Proc. Natl. Acad. Sci. USA* **98:** 5116–5121.
9. Kanehisa M and Goto S. (2000) KEGG: Kyoto encyclopedia of genes and genomes. *Nucleic Acids Res.* **28:** 27–30.
10. Falcon S and Gentleman R. (2007) Using GOstats to test gene lists for GO term association. *Bioinformatics* **23:** 257–258.

9. Dynamic Modeling

Yu-Chao Wang

Institute of Biomedical Informatics, National Yang-Ming University, Taiwan
yuchao@ym.edu.tw

1. Introduction

Biological modeling is the activity of translating a biological system into mathematics for subsequent analysis. In the field of systems biology, the end goal of biological modeling is to understand the interactions between the components of the system and to have a fully predictive description of the system, that is, computational simulations are guaranteed to be accurate representations of real behavior of the biological system. To this end, differential equations are the most common mathematical tools to model biological systems. Briefly, mathematical models can be classified into two categories: static models where the variables of the system do not change in time, and dynamic models which account for time-dependent changes of the system.[1] Since most biological systems are time-dependent and the dynamic behaviors are crucial for understanding biological mechanisms, we will focus on the dynamic models in this chapter. In the previous chapters, the generation and analysis of high-throughput biological data such as transcriptomics, quantitative proteomics, and phosphoproteomics are introduced. Here, we will introduce the procedure of dynamic modeling and the use of dynamic modeling to analyze those high-throughput data for understanding the interactions/regulations between the components of biological systems, i.e., constructing the biological networks.

2. Mathematical fundamentals

2.1. *Introduction to differential equations*

A differential equation is a mathematical equation that contains derivatives. Basically, the differential equation expresses the rate of change of a quantity (the state variable) as a function of time or the state variable itself. For example,

$$\frac{dm}{dt} = f(m,t) \tag{9.1}$$

where m is the state variable, t denotes the time, $f(m,t)$ represents a function of time and state variable m, and $\frac{dm}{dt}$ indicates the rate of change of state variable m. In systems biology, the state variables are usually represented by the expression level of genes, proteins, or phosphoproteins.

   Given the differential equation, we always want to know the solution of the differential equation, that is, the quantity of state variable over time. When solving the differential equation, we need to have the initial condition of the state variable, i.e., the value of the state variable at any given time. With the differential equation and the initial condition of the state variable, calculus techniques can be applied to analytically solve the differential equation. However, not all differential equations can be solved analytically by calculus techniques. In this case, the Euler method and the Runge–Kutta method are two popular ways to numerically solve the differential equations,[2] which are useful for computational simulations.

2.2. *From differential equations to discrete dynamic models*

Traditionally, differential equations are used to describe continuous-time systems. Although biological systems are mostly continuous-time systems, the experimental data generated from the biological systems such as transcriptomics, quantitative proteomics, and phosphoproteomics data are always discrete-time. In this case, it is not appropriate to fit discrete-time experimental data to differential equations. Consequently, the differential equations need to be modified as the discrete dynamic models. Based on the definition of the derivative, the differential equation in (9.1) can be expressed as

$$\frac{m(t+\Delta t) - m(t)}{\Delta t} = f(m,t) \tag{9.2}$$

By transposition, Eq. (9.2) can be further expressed as

$$m(t + \Delta t) = m(t) + f(m, t)\Delta t$$
$$= g(m, t) \tag{9.3}$$

which becomes a discrete dynamic model (or difference equation). The physical meaning of the discrete dynamic model is that the value of the state variable m at time $t + \Delta t$ is a function of the state variable m at time t. In other words, given the value of state variable m at time t and the discrete dynamic model (9.3), the value of the state variable m at time $t + \Delta t$ can always be computed. In the following sections, the discrete dynamic model will be used to describe the interactions/regulations between the components of the biological system.

3.   The modeling procedure

In this section, the step-by-step modeling procedure will be introduced in detail. Here, the analysis of transcriptomics data using dynamic model to investigate the regulations among genes is taken as an example to demonstrate the modeling procedure.

3.1.   *Problem formulation and key factor determination*

The first step of the modeling procedure is to define the problem and to set the goal based on experimental data we have. Suppose that we have a set of transcriptomics data which measured the time-course gene expression profiles under a specific experimental condition, e.g., some kind of drug treatment. The goal of the analysis is to investigate gene regulations between the genes of interest, i.e., the construction of gene regulatory network for the specific experimental condition. In this situation, how to determine the gene regulation relationships based on the transcriptomics data is our main concern.

There are thousands of genes in an organism, indicating that it is not practical to investigate gene regulations among all genes. Therefore, we need to determine the genes of interest for further analysis. Generally, the selection of genes can be divided into two categories: expression-based selection and function-based selection. For expression-based selection, statistical methods such as one-way analysis of variance (ANOVA) or simply fold change selection are usually applied to gene expression profiles from the transcriptomics data for global selection of genes of interest. In this case, the constructed network will represent the global scenario for all the dynamically regulated genes under the experimental condition. If one-way ANOVA is employed to detect significant gene expression variations across different time-points for each gene, the null hypothesis of ANOVA assumes that the average expression level of a gene would be the same at every time point.[3] Usually, genes with Bonferroni-adjusted p-values of less than 0.05 are identified as dynamically regulated genes and selected as the genes of interest. On the other hand, the function-based selection method is applied only if we want to construct the network for some specific functions. Gene ontology annotations are useful tools for functional annotation of genes.[4]

3.2. *Model structure construction*

The construction of model structure represents determination of the updating function $g(m, t)$ (the right-hand side of the discrete dynamic model (9.3)), which depends on the prior knowledge about the investigated biological systems and the experimental data itself. In the case of gene regulation modeling, we consider a gene regulation relationship as a system block with several regulatory genes as inputs and a target gene as output. Based on our knowledge about the transcriptional regulation, for a target gene i among the genes of interest, the temporal expression of the target gene can be described by the following discrete dynamic model:

$$x_i(t+1) = x_i(t) + \sum_{j=1}^{N_i} a_{ij}x_j(t) - \lambda_i x_i(t) + k_i + \varepsilon_i(t) \qquad (9.4)$$

where $x_i(t)$ represents the gene expression level of target gene i at time t, a_{ij} denotes the regulatory ability of the j-th regulatory gene for the target gene i with a positive value indicating activation and a negative value indicating repression, $x_j(t)$ represents the gene expression level for the j-th regulatory gene that potentially regulates target gene i, N_i denotes the number of genes potentially regulating gene i, λ_i indicates the degradation effect of the target gene i, k_i represents the basal expression level, and $\varepsilon_i(t)$ represents the stochastic noise due to the model uncertainty. The biological implication of the discrete dynamic model (9.4) is that the gene expression level of target gene i at the next time $t + 1$ is determined by the present gene expression level, the regulation of N_i regulatory genes, the degradation effect of the present time, the basal level of gene expression, and some stochastic noises,[5] which coincides with our realization of transcriptional regulation.

In the discrete dynamic model of gene regulation (9.4), the regulatory genes that potentially regulates target gene i could be all genes or some subset of genes among the genes of interest. Some other kinds of data such as ChIP-on-chip (ChIP-chip) or ChIP-sequencing (ChIP-seq) data demonstrating the protein-DNA interaction information may be integrated to screen the potential genes that regulates target gene i. In that case, the number of genes potentially regulating gene i, N_i, would be different for each gene. Furthermore, in addition to the state variables $(x_i(t), x_j(t)$'s) in (9.4), there are many model parameters, i.e., a_{ij}'s, λ_i, k_i, which characterize the interactions among the system components and environmental effects. Since the main objective is to investigate gene regulations between the genes of interest, the regulatory abilities a_{ij}'s, whose values quantify the regulatory strength between gene i and gene j, are of special importance. These model parameters would be identified with the aid of experimental data in the next section.

3.3. *Model parameter identification*

Following formation of the discrete dynamic models describing temporal expression of the selected genes of interest, the parameters in the models require identification using the transcriptomics data. The model parameter identification means adjusting the parameters of the model until the behavior of the model matches the generated experimental data. Maximum likelihood estimation method is one of the frequently used methods for parameter identification,[6] which is introduced here. The discrete dynamic model (9.4) can be written in the following regression form:

$$x_i(t+1) = [x_1(t) \quad \cdots \quad x_{N_i}(t) \quad x_i(t) \quad 1] \cdot \begin{bmatrix} a_{i1} \\ \vdots \\ a_{iN_i} \\ (1-\lambda_i) \\ k_i \end{bmatrix} + \varepsilon_i(t)$$

$$\equiv \phi_i(t) \cdot \theta_i + \varepsilon_i(t) \tag{9.5}$$

where $\phi_i(t)$ denotes the regression vector which can be obtained from the transcriptomics data, and θ_i is the parameter vector to be estimated for target gene i. In order to avoid the danger of overfitting the estimated parameters, the original data points are interpolated to L data points by the cubic spline method[7] (L should be larger than the number of parameters to be estimated). In other words, there are $\{x_i(l+1), \phi_i(l)\}$ data point pairs for

$l \in \{1, \ldots, L - 1\}$. Hence, Eq. (9.5) can be written in the following matrix form for target gene i:

$$X_i = \Phi_i \cdot \theta_i + E_i \tag{9.6}$$

where

$$X_i = \begin{bmatrix} x_i(2) \\ \vdots \\ x_i(L) \end{bmatrix}, \quad \Phi_i = \begin{bmatrix} \phi_i(1) \\ \vdots \\ \phi_i(L-1) \end{bmatrix}, \quad E_i = \begin{bmatrix} \varepsilon_i(1) \\ \vdots \\ \varepsilon_i(L-1) \end{bmatrix}$$

In Eq. (9.6), we assume noises $\varepsilon_i(l)$ at different time points as independent random variables of normal distribution with zero mean and unknown variance σ_i^2, i.e., the variance of ε_i is $\Sigma_i = E\{\varepsilon_i \varepsilon_i^T\} = \sigma_i^2 I$, where I is an identity matrix. If ε_i is assumed to be normally distributed with $L - 1$ elements, its probability density function is of the following form[6]:

$$p(\varepsilon_i) = \left((2\pi)^{L-1} \det \Sigma_i\right)^{-1/2} \exp\left\{-\frac{1}{2}\varepsilon_i^T \Sigma_i^{-1} \varepsilon_i\right\} \tag{9.7}$$

Considering Eqs. (9.6) and (9.7), the likelihood function can be expressed as

$$L(\theta_i, \sigma_i^2) = p(\theta_i, \sigma_i^2)$$
$$= (2\pi\sigma_i^2)^{-(L-1)/2} \exp\left\{-\frac{1}{2\sigma_i^2}(X_i - \Phi_i\theta_i)^T(X_i - \Phi_i\theta_i)\right\} \tag{9.8}$$

Maximum likelihood estimation method aims at finding θ_i and σ_i^2 to maximize the likelihood function in Eq. (9.8). For the simplicity of computation, it is practical to take the logarithm of the likelihood function, and we have the following log-likelihood function:

$$\ln L(\theta_i, \sigma_i^2) = -\frac{L-1}{2}\ln(2\pi\sigma_i^2) - \frac{1}{2\sigma_i^2}\sum_{l=1}^{L-1}(x_i(l+1) - \phi_i(l) \cdot \theta_i)^2 \tag{9.9}$$

where $x_i(l+1)$ and $\phi_i(l)$ are the l-th elements of X_i and Φ_i, respectively. Here, the log-likelihood function is expected to have the maximum at $\theta_i = \hat{\theta}_i$ and $\sigma_i^2 = \hat{\sigma}_i^2$. The necessary conditions for determining the maximum likelihood estimates $\hat{\theta}_i$ and $\hat{\sigma}_i^2$ must conform to the following two equations:

$$\frac{\partial \ln L(\theta_i, \sigma_i^2)}{\partial \theta_i}\bigg|_{\theta_i = \hat{\theta}_i} = 0$$

$$\frac{\partial \ln L(\theta_i, \sigma_i^2)}{\partial \sigma_i^2}\bigg|_{\sigma_i^2 = \hat{\sigma}_i^2} = 0$$

(9.10)

After some computational deduction, the estimated parameters $\hat{\theta}_i$ and $\hat{\sigma}_i^2$ are given as

$$\hat{\theta}_i = (\Phi_i^T \Phi_i)^{-1} \Phi_i^T X_i$$

(9.11)

$$\hat{\sigma}_i^2 = \frac{1}{L-1} \sum_{l=1}^{L-1} (x_i(l+1) - \phi_i(l) \cdot \hat{\theta}_i)^2$$

$$= \frac{1}{L-1} (X_i - \Phi_i \hat{\theta}_i)^T (X_i - \Phi_i \hat{\theta}_i)$$

(9.12)

In this manner, given the transcriptomics data, the parameters in the discrete dynamic model can be identified based on Eq. (9.11) Simply speaking, regardless of all the mathematic deduction, we can use the transcriptomics data to build the matrices Φ_i and X_i in (9.6) according to Eq. (9.5). Then, with the formula in (9.11), simple matrix manipulation can be employed to have the estimated parameter vector $\hat{\theta}_i$, from which all the model parameters can be identified.

3.4. *Model selection*

In the discrete dynamic model describing the temporal expression of gene, the parameter a_{ij} denotes the regulatory ability of the j-th regulatory gene for the target gene i, which implies the regulatory relationships between the genes. Since the objective is to identify the mechanism of regulatory relationships of the selected genes of interest, the significance of the regulatory abilities should be determined for the identification of significant regulations in the gene regulatory network. Therefore, Akaike information criterion (AIC)[6,8] and Student's t-test[3] are then employed for model order selection and to determine the significance of the regulatory relationships. AIC, which includes both estimated residual error and model complexity in one statistics, quantifies the relative goodness of fit of a model. For a discrete dynamic model with N_i parameters (or regulatory genes) to fit with data from L samples, the AIC statistics can be written as follows[6,8]:

$$\text{AIC}(N_i) = \log\left(\frac{1}{L}(X_i - \hat{X}_i)^T(X_i - \hat{X}_i)\right) + \frac{2N_i}{L} \tag{9.13}$$

where $\hat{X}_i$ denotes the estimated expression profile of the i-th target gene, i.e. $\hat{X}_i = \Phi_i \cdot \hat{\theta}_i$, and $\hat{\sigma}_i^2 = \frac{1}{L}(X_i - \hat{X}_i)^T(X_i - \hat{X}_i)$ is the estimated residual error. As the residual error $\hat{\sigma}_i^2$ decreases, the AIC decreases. In contrast, while the number of regulatory genes (or parameters) N_i increases, the AIC increases. Therefore, there is a tradeoff between residual error and model order. As the expected residual error decreases with increasing number of regulatory genes in models of inadequate complexity, there should be a minimum around the optimal number of regulatory genes. The minimization achieved in Eq. (9.13) will indicate the ideal model order (i.e. the optimal number of gene that regulate the target gene) of the discrete dynamic model. With the statistical selection of N_i regulatory genes by minimization of the AIC, the question of whether a regulatory gene is a significant one or just a false positive for the i-th target gene can be determined. Hence, AIC can be adopted to select model order, filtering out insignificant regulations in the gene regulatory network based on the estimated regulatory abilities (a_{ij}'s). Due to computational efficiency, it is impractical to compute the AIC statistics for all possible regression models. Stepwise methods such as forward selection method and backward elimination method are developed to avoid the complexity of exhaustive search.[9,10] However, in the case of backward selection method, a variable once eliminated can never be reintroduced into the model, and in the case of forward selection, once included can never be removed.[10] Thus, the stepwise regression method which combines forward selection method and backward elimination method is suggested to be used to compute the AIC statistics. In addition to AIC model selection criteria, the Student's t-test[3] is further employed to calculate the p-values for the regulatory abilities under the null hypothesis $H_0 : a_{ij} = 0$ to determine the significant regulatory relationships. The regulations with p-value ≤ 0.05 are determined as significant regulations and preserved in the gene regulatory network. According to these modeling procedures, the gene regulatory network for the specific experimental condition is constructed from the transcriptomics data.

4. Summary and perspective

In this chapter, we demonstrate the use of dynamic modeling techniques to analyze high-throughput data for understanding the interactions/regulations between the components of biological systems. Although the analysis of transcriptomics data for constructing the gene regulatory network is taken as the example, other high-throughput data such as quantitative proteomics and phosphoproteomics data can also be analyzed with the similar procedure. The analysis of high-throughput data with dynamic modeling procedure results in the mathematical expression of the biological systems, which will be valuable for computational simulation of the systems. In addition, the biological networks under specific experimental conditions can be constructed based on the modeling procedure. These networks can be further analyzed for understanding the biological mechanisms. For example, the network motifs among the biological networks can be identified for providing the insights into the network's functions.[11] Network modules can also be determined to predict unknown gene function, prioritize disease genes, and classify cancer subtypes, etc.[12] Furthermore, biological networks under different experimental conditions can be compared to become the differential networks, which can yield insight into the biological basis of variations of different phenotypes or diseases.[13] In summary, the dynamic modeling procedure introduced in this chapter provides a useful tool for high-throughput data analyses in systems biology.

References

1. Sauro HM. (2014) *Essentials of Biochemical Modeling.* Ambrosius Publishing.
2. Adler FR. (2013) *Modeling the Dynamics of Life: Calculus and Probability for Life Scientists*, 3rd Edn. Brooks/Cole, Cengage Learning.
3. Pagano M and Gauvreau K. (2000) *Principles of Biostatistics*, 2nd edn. Duxbury.
4. Ashburner M, Ball CA, Blake JA *et al.* (2000) Gene ontology: Tool for the unification of biology. The Gene Ontology Consortium. *Nat. Genet.* **25**: 25–29.
5. Wang YC and Chen BS. (2010) Integrated cellular network of transcription regulations and protein–protein interactions. *BMC Syst. Biol.* **4**: 20.
6. Johansson R. (1993) *System Modeling and Identification*, Prentice Hall.
7. Faires JD and Burden RL. (2013) *Numerical Methods*, 4th edn. Brooks/Cole, Cengage Learning.
8. Akaike H. (1974) A new look at the statistical model identification. *IEEE Trans. Autom. Contr.* **19**: 716–723.
9. Hocking RR. (1976) A Biometrics invited paper. The analysis and selection of variables in linear regression. *Biometrics* **32**: 1–49.
10. Seber GAF and Lee AJ. (2003) *Linear Regression Analysis*, 2nd edn. Wiley-Interscience.
11. Alon U. (2007) *An Introduction to Systems Biology: Design Principles of Biological Circuits.* Chapman & Hall/CRC.
12. Mitra K, Carvunis AR, Ramesh SK *et al.* (2013) Integrative approaches for finding modular structure in biological networks. *Nat. Rev. Genet.* **14**: 719–732.
13. Ideker T and Krogan NJ. (2012) Differential network biology. *Mol. Syst. Biol.* **8**: 565.

10. Protein Structure Modeling

Chia-Hsien Lee and Hsueh-Fen Juan*

Graduate Institute of Biomedical Electronics and Bioinformatics
National Taiwan University, Taipei, Taiwan
*yukijuan@ntu.edu.tw

1. Introduction

Protein structure plays important roles in various parts of biological research. At first, researchers obtain protein structure experimentally with X-ray diffraction to learn mechanism of action of their protein or biochemical reaction steps of the enzyme, or the reason why some mutations bring disaster on protein function. In summary, we learned a lot from those protein structures, and thanks to them, our knowledge in biochemistry and certain other biology disciplines have broadened.

Protein structure is now an important resource in computer-aided drug discovery. For example, if one has a protein structure, he may use appropriate algorithm to deduce how a compound (or ligand) binds to the protein and evaluates the effectiveness of the ligand as a drug. Using this method we developed a lot of drugs for many diseases. For example, in the work of Harikishore et al.,[1] crystal structure of hepatitis C virus NS5B is utilized to find novel drugs targeting allosteric sites with docking and pharmacophore techniques.

Despite the importance of protein structure information, obtaining them experimentally is a tedious job. It may take many years to successfully obtain crystallized pure protein which is suitable for X-Ray diffraction experiment. Furthermore, for some proteins, it's hard to obtain protein crystals of acceptable quality. For example, membrane proteins are long considered hard-to-form crystal; however, it may be far easier due to recent advance on related technologies.[2] Fortunately, some researchers developed methods to deduce protein structure computationally, trying to solve such problems.

Currently, methods of computational protein structure deduction can be separated into two categories: *ab initio* (which is "from the beginning" in Latin) and homology modeling. Methods classed as *ab initio* methods will take protein as their sole input, model protein structure with knowledge of physics and biochemistry from scratch by using techniques such as molecular dynamics, and output the model. In the work of Chowdhury *et al.*,[3] part of miniprotein tc5b is folded by using molecular dynamics, and the result highly resembles the structure from nuclear magnetic resonance (NMR). However, as for now, *ab initio* methods are not reliable, and cannot easily work in large proteins since they will consume too much computational resources.[4,5] In the second category, homology modeling, a protein model is created from similar proteins with structures. That is, query sequence is compared with other similar sequence(s) with structures. The homology modeler will then assemble protein models according to the similarity. The more identical residual exists between the template and the query, according to their sequence alignment, the more accurate produced structure is.[6] In this chapter, we focus on homology modeling (Fig. 1), since it's the most common technique used in protein modeling, and introduce some useful tools used in homology modeling procedure.

In homology modeling, first, one should obtain protein sequence of the target protein from a protein database, such as UniProt[7] or NCBI Protein.[8] Second, an appropriate template should be selected from structure database like Protein Data Bank and alignment of query sequence and template should be created. Then, the alignment will be submitted to the modeling software such as SwissModel,[9] I-tasser,[10] or ModWeb.[11] In some modeling software package, template search and modeling are automatically done without user interference. However, before running homology modeling, some model databases, for example, ModBase[12] and Protein Model Portal,[13] can be searched for models created by other researchers. Finally, to check if quality of the models is acceptable, protein quality assessment tools can be applied on the models. ModEval[14] and Qmean server[15] can do protein quality assessment for those who want to check their protein structure models.

1.   Checking existing computational protein models

Before you dive into homology modeling, checking if existing model exists is recommended, since homology modeling is a time-consuming process. Such checking may prevent unnecessary resource waste.

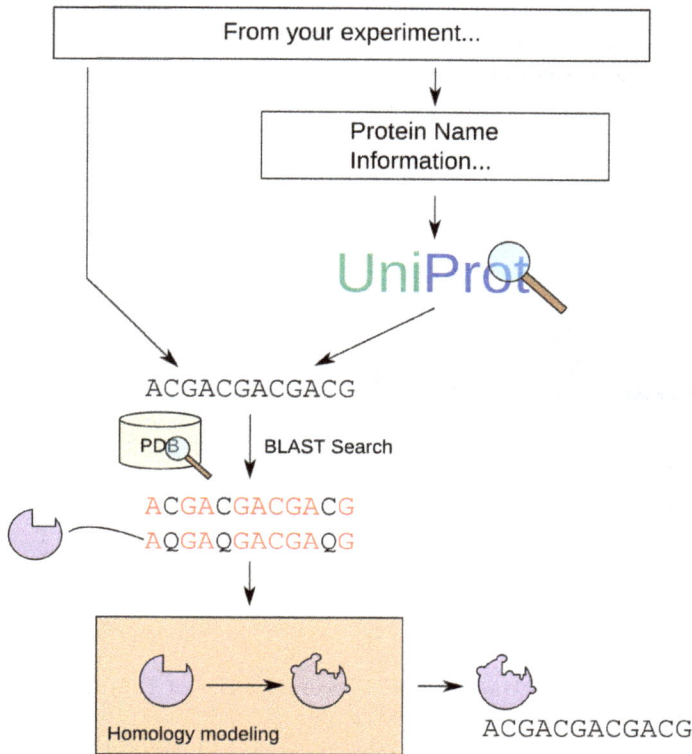

Figure 1. General step(s) used in homology modeling. First, one should obtain protein sequence from a protein database with pre-acquired knowledge of that protein if they don't have the sequence of protein of interest. Second, an appropriate template should be obtained by search protein structure database. Third, an alignment should be generated between the template and the protein of interest. Finally, the alignment is used as input of a homology modeling application to generate the final model structure.

Materials

- Information about your protein. For example, common name and gene symbol.

Procedure

I. Navigate to UniProt: http://www.uniprot.org/, or search UniProt in the search engine (Fig. 2(a)).

II. Input name or gene symbol of your protein in the search bar. Use advanced search option (on the right side of the search bar) when appropriate.

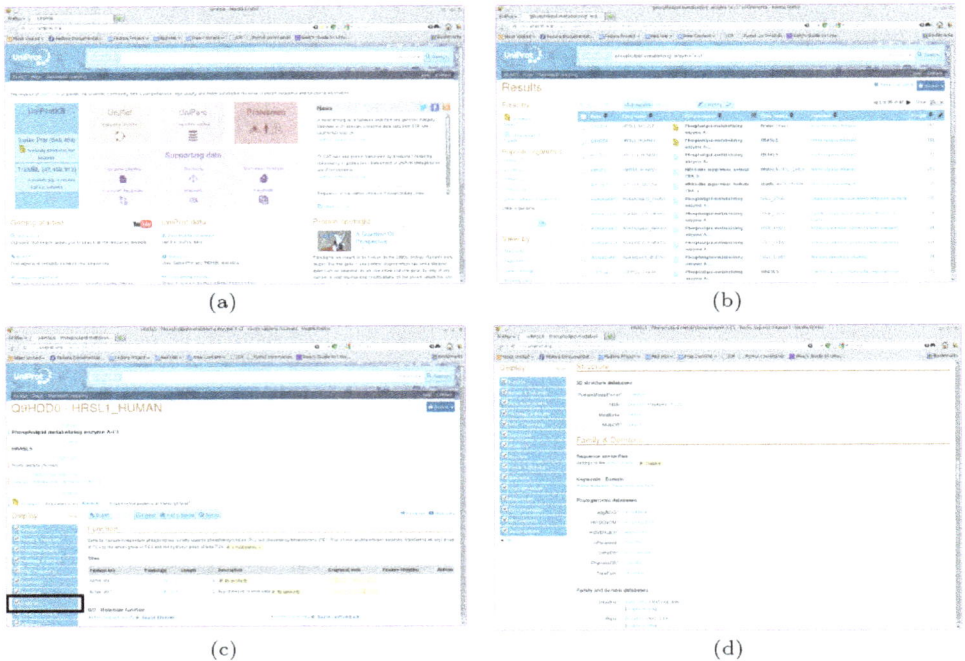

(a)

(b)

(c)

(d)

Figure 2. Screenshots of UniProt. (a) Welcome screen of UniProt. You can see the search bar in the top of the web page, and 'advanced' and 'search' buttons on the right of the search bar. (b) The result list given after search. (c) Protein information page. In the black box mark is the 'structure' navigation button. (d) Structure-related information of selected protein.

III. From the given search result list, pick the one(s) that match your protein and click its (their) accession number to enter its (their) page (Fig. 2(b)).

IV. Inspect information given by UniProt to make sure it is your selected protein. If it is confirmed to be your protein of interest, take note of its accession number for further use (Fig. 2(c)).

V. Navigate to 'Structure' section by clicking 'Structure' button on the left panel (Fig. 2(c)).

VI. Here, you may see one or more links to Protein Data Banks if experimental structure exists. You can see some links to ModBase and Protein Model Portal. You may click on those links and check if the structures provided in those sites are acceptable for your standard. You may also check quality of the models with the quality check server listed below (in Secs. 5 and 6, Fig. 2(d)).

2. Homology modeling with SwissModel

You may choose SwissModel, ModWeb, or I-tasser, or all of them to do homology modeling. Difference between SwissModel and I-tasser are described in 'I-tasser' section.

Materials

- Accession number of your protein in UniProt database, or sequence of your protein of interest. To get accession number of a protein, please check the previous section.

Procedure

I. Navigate to UniProt (http://www.uniprot.org/), or search UniProt in the search engine (If you possess your protein sequence, please skip to step VI. If you wish to give up the opportunity of downloading and saving protein sequence while you have its accession number, you may proceed to step VI too.)

II. Input accession number into the search bar and hit 'search'.

III. Check the information shown on the page to ensure that the information on that page is about your protein before you proceed.

IV. Navigate to 'Sequence section by clicking 'Section' button on the left panel.

V. Click on the 'FASTA' button above the box displaying your sequence and copy the FASTA sequence shown. If there are multiple isoforms, please check all of them and pick one matching your need. It is recommended to save the sequence along with the model created by SwissModel to your hard drive together (Fig. 3(a)).

VI. Navigate to SwissModel (http://swissmodel.expasy.org/) or search it in the search engine. Hit 'start modeling' button on the welcome page (Fig. 3(b)).

VII. Paste the protein sequence or UniProt accession number into 'Target Sequence' field. Fill others accordingly. For other acceptable input format, please refer to the right panel of this screen. It is highly recommended that you fill 'email' field so that the result will be mailed to you inbox after the modeling is completed.

If something goes wrong while you input target sequence, press 'Reset Form' button and do it again.

(a)   (b)

(c)

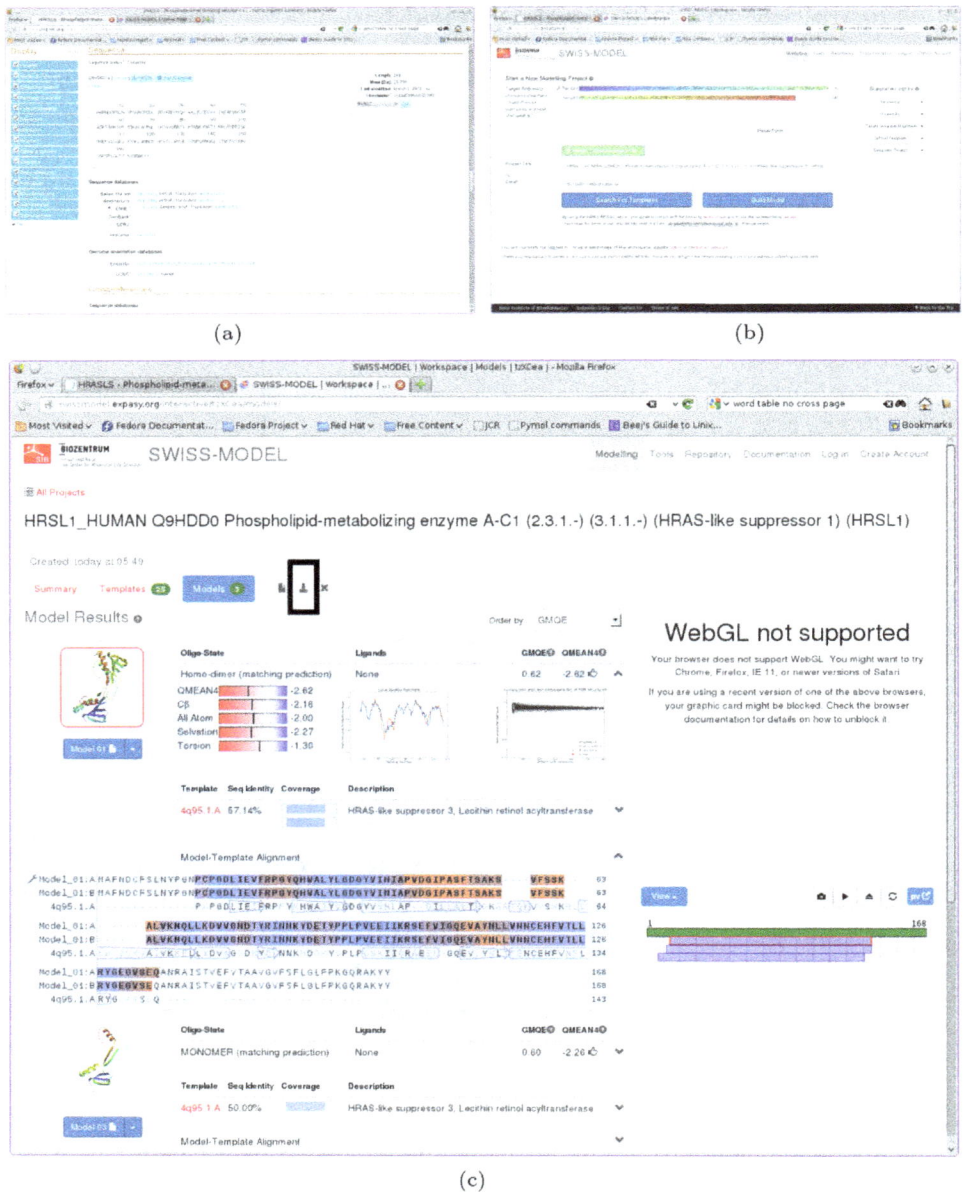

Figure 3.   Using SwissModel for protein modeling. (a) In 'sequence' section of protein information page of UniProt, you can view and download its sequence for use. (b) Swiss-Model job submitting page. (c) Result page given by SwissModel. You can assess the quality of the model here. The whole result dataset can be downloaded with the download button (marked with black box).

VIII. Read the text below 'Search for Templates'. If you accept the term below, hit 'Search for Templates' to begin structure modeling.

If you do not want to participate in template searching process, press 'Build Model' instead and go to step X.

IX. After template searching is completed, inspect the list and choose the best one for your situation. You may click on the 'v' icon in the end of the row for some more information of the alignment. After you select the best template, click the 'v' icon in the end of the row and hit 'Build Model' button to make the model. Please DO NOT close your browser if you didn't fill your email when submitting your modeling job.

X. When the modeling is completed, you may navigate to the result screen, inspect the report about information of the models and beautifully-drawn figures about the models. You may download the parts you need, or download everything by clicking on the 'download' icon, which is under your project title and the right side of 'Model' tab (Fig. 3(c)).

Downloading everything and saving them to your disk is highly recommended.

3. Homology modeling with I-tasser web

There are some differences between I-tasser and SwissModel.

First, SwissModel will model the part with acceptable templates, thus some part of the heading and tailing sequence of your input might not be modeled and cannot be found in your resulting model. On the other hand, I-tasser will strive to model the whole protein for you. In almost all cases, everything in your input sequence will be presented in the model which is produced by I-tasser.

Second, I-tasser will try to do function and ligand prediction for the best model of its output. This might be helpful if known information of this protein is too few.

Materials

• Sequence of your protein of interest. To get sequence of a protein, please check the previous sections.

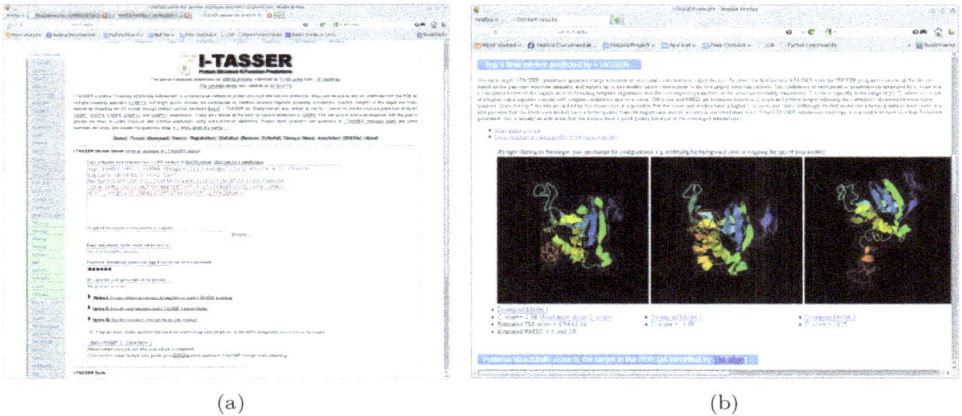

(a)   (b)

Figure 4.   Modeling with I-tasser Web. (a) Job submission form of I-tasser. (b) Result page (partial) of I-tasser. You may download the result dataset, packed into a compressed file, with the link provided in the top of the report (not shown).

Procedure

I. Navigate to I-tasser (http://zhanglab.ccmb.med.umich.edu/I-TASSER/), or search it in the search engine (Fig. 4(a)).

II. Paste the protein sequence into the large input box, or upload the sequence file by using the file upload field below. Fill the other fields if found necessary. Entering you e-mail is recommended (Fig. 4(a)).

You may also view the options below to adjust them if needed. In most cases, the defaults are good to go.

You may request an account and password for tracking your job progress.

III. Click 'Run I-tasser' below. Depending on properties of your protein, it may take at most three to four days to finish the work.

IV. After the modeling is finished, you may navigate the result page and view information about template, scoring, function prediction, and ligand prediction information in this page (Fig. 4(b)).

To download everything, please use the link directly below title of the result page.

4. Homology modeling with ModWeb

You may use ModWeb (Fig. 5), which uses Modeler as its modeling engine. Using ModWeb requires a MODELLER license key, which can be obtained freely for academic users in http://salilab.org/modeller/registration.html.

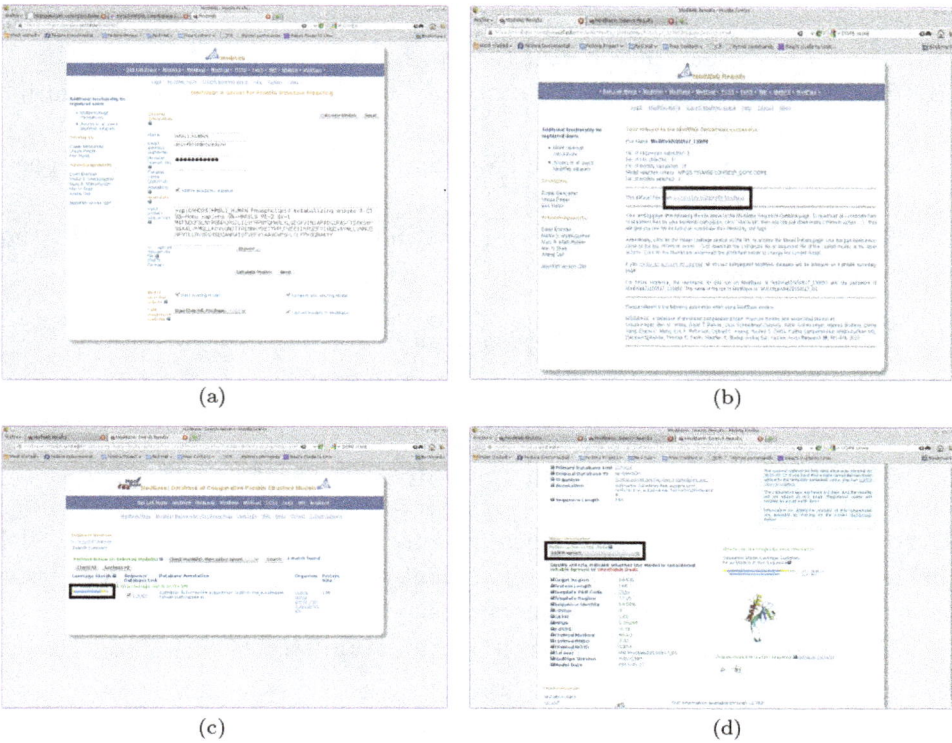

(a)

(b)

(c)

(d)

Figure 5. Modeling with ModWeb. (a) ModWeb job submitting form. (b) Result page, notifying you to view the result in ModBase page (by link mark by black box), where your result is stored. (c) Overview of the models generated. Click on the bar marked by the black box to go to the detailed information page. (d) Page showing detailed information about your model. Click on the combo box and select 'Coordinate File' to download the structure file.

Materials

- MODELLER License key
- Sequence of your protein of interest. To get sequence of a protein, please check the previous sections.

Procedure

I. Navigate to ModWeb (https://modbase.compbio.ucsf.edu/modweb/), or search it in the search engine.
II. Fill the name, input sequence of the protein of interest, and license key into the form. Fill other things if appropriate (Fig. 5(a)).

It's highly recommended that you fill in your e-mail address into the form, since it may take a long time to finish your model.

III. Press 'Calculate Model'. Take note of the link in the result page for future reference.

IV. After the model is finished, you may navigate to the result page (Fig. 5(b)).

 V. Navigate to the detail page as show in Figs. 5(b–d) and inspect the result carefully.

Note: please save the web page containing detailed description of your mode along with your model coordinate file (i.e., pdb file)

5. Quality assessment with QMean server (Fig. 6)

Although the modeling report given by I-tasser, ModWeb and SwissModel contains quality assessment, you might want to gain more confidence of your model by submitting the model for 3rd party inspection.

QMean score is already used in the evaluation step of SwissModel.

Materials

• The model created by I-tasser or ModWeb. (Since models from SwissModel uses the same scoring matrix as QMean server, submitting them would be unnecessary)

Procedure

 I. Navigate to QMean server page by entering its url into your browser's navigation bar (http://swissmodel.expasy.org/qmean/cgi/index.cgi) or search 'QMean server' in your search engine (Fig. 6(a)).

II. Fill the form accordingly. Hit 'Submit' to start the job (Fig. 6(a)).

III. After the model evaluation is finished, you may navigate to evaluation report (Fig. 6(b)).

You can see global score and local score detail from the result page.

To get more information about Q-MEAN score and Q-MEAN Z-score, please press the question mark buttons, or click on 'help' button in the navigation bar under the title 'QMEAN Server for Model Quality Estimation'.

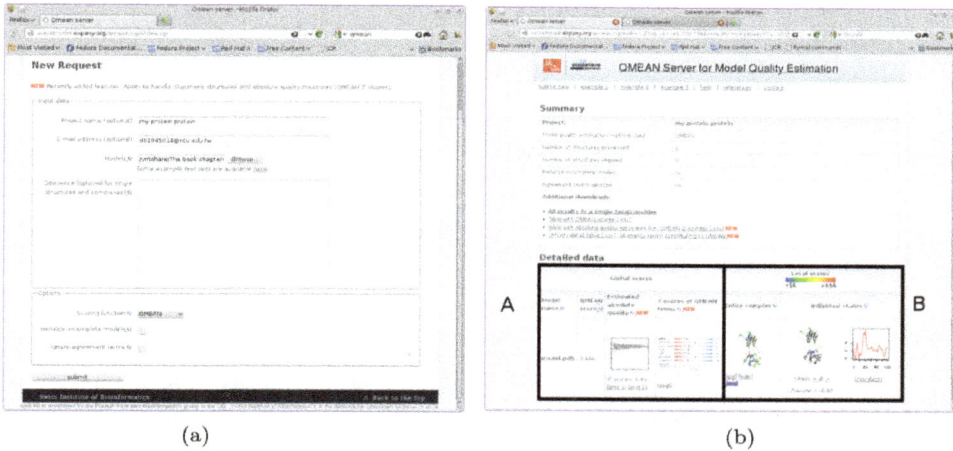

(a)   (b)

Figure 6. Quality assessment with QMean server. (a) Job submission page and (b) evaluation report pertaining information about overall quality of the model (black box A) and local quality of the model (black box B).

6. Quality assessment with ModEval sever

ModEval will give you the same scores as the ones used in ModWeb.

Materials

- The model created by I-tasser or SwissModel. (Since models from ModWeb use the same scoring matrix as ModEval server, submitting them would be unnecessary)
- Sequence identity between your query sequence and template sequence. For models obtained from SwissModel, open the model pdb file with a text editor, search for "SID" in 'REMARK' section (Fig. 7(a)). For models obtained from ModWeb, open the model pdb file with a text editor and search for "SEQUENCE IDENTITY" (Fig. 7(b)).

> Beware that ModEval require an integer as sequence identity input.
>
> Since I-tasser uses multiple templates in the process of model construction, ModEval may not be suitable for evaluating I-tasser models.

- MODELLER License key. See 'ModWeb' section above for method of obtaining license key.

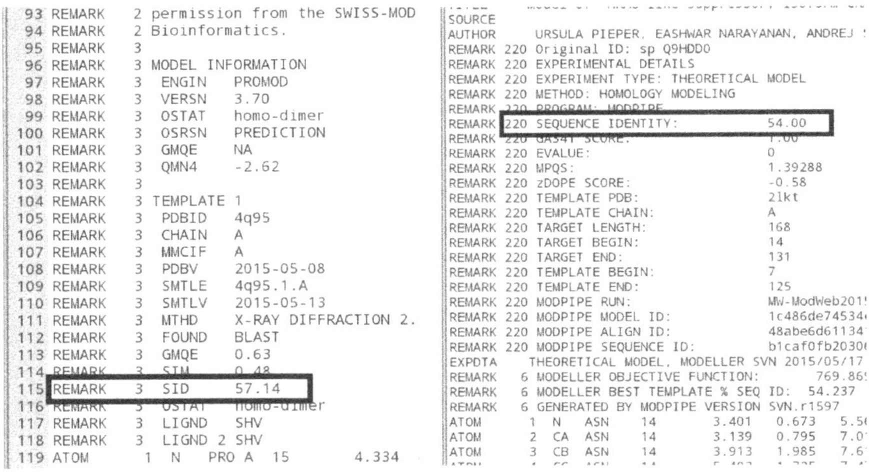

```
 93 REMARK    2 permission from the SWISS-MOD          SOURCE
 94 REMARK    2 Bioinformatics.                        AUTHOR     URSULA PIEPER, EASHWAR NARAYANAN, ANDREJ
 95 REMARK    3                                        REMARK 220 Original ID: sp Q9HDD0
 96 REMARK    3 MODEL INFORMATION                      REMARK 220 EXPERIMENTAL DETAILS
 97 REMARK    3   ENGIN    PROMOD                       REMARK 220 EXPERIMENT TYPE: THEORETICAL MODEL
 98 REMARK    3   VERSN    3.70                         REMARK 220 METHOD: HOMOLOGY MODELING
 99 REMARK    3   OSTAT    homo-dimer                   REMARK 220 PROGRAM: MODPIPE
100 REMARK    3   OSRSN    PREDICTION                   REMARK 220 SEQUENCE IDENTITY:        54.00
101 REMARK    3   GMQE     NA                           REMARK 220 GA341 SCORE:              1.00
102 REMARK    3   QMN4     -2.62                        REMARK 220 EVALUE:                   0
103 REMARK    3                                        REMARK 220 MPQS:                   1.39288
104 REMARK    3 TEMPLATE 1                              REMARK 220 zDOPE SCORE:            -0.58
105 REMARK    3   PDBID    4q95                         REMARK 220 TEMPLATE PDB:           2lkt
106 REMARK    3   CHAIN    A                            REMARK 220 TEMPLATE CHAIN:         A
107 REMARK    3   MMCIF    A                            REMARK 220 TARGET LENGTH:          168
108 REMARK    3   PDBV     2015-05-08                   REMARK 220 TARGET BEGIN:           14
109 REMARK    3   SMTLE    4q95.1.A                     REMARK 220 TARGET END:             131
110 REMARK    3   SMTLV    2015-05-13                   REMARK 220 TEMPLATE BEGIN:         7
111 REMARK    3   MTHD     X-RAY DIFFRACTION 2.         REMARK 220 TEMPLATE END:           125
112 REMARK    3   FOUND    BLAST                        REMARK 220 MODPIPE RUN:            MW-ModWeb201!
113 REMARK    3   GMQE     0.63                         REMARK 220 MODPIPE MODEL ID:       1c486de74534
114 REMARK    3   SIM      0.48                         REMARK 220 MODPIPE ALIGN ID:       48abe6d61134
115 REMARK    3   SID      57.14                        REMARK 220 MODPIPE SEQUENCE ID:    b1caf0fb2030(
116 REMARK    3   OSTAT    homo-dimer                   EXPDTA     THEORETICAL MODEL, MODELLER SVN 2015/05/17
117 REMARK    3   LIGND    SHV                          REMARK   6 MODELLER OBJECTIVE FUNCTION:       769.86!
118 REMARK    3   LIGND 2 SHV                           REMARK   6 MODELLER BEST TEMPLATE % SEQ ID:   54.237
119 ATOM      1   N   PRO A  15        4.334            REMARK   6 GENERATED BY MODPIPE VERSION SVN.r1597
                                                        ATOM      1  N   ASN   14        3.401   0.673   5.5(
                                                        ATOM      2  CA  ASN   14        3.139   0.795   7.0'
                                                        ATOM      3  CB  ASN   14        3.913   1.985   7.6'
```

              (a)   (b)

Figure 7.   Sequence identity information (marked by black box) can be obtained from the result models of (a) SwissModel and (b) ModWeb.

Procedure

   I. Navigate to ModEval page by entering its url into your browser's navigation bar (http://modbase.compbio.ucsf.edu/evaluation/) or search 'ModEval' in your search engine (Fig. 8(a)).
  II. Fill the form accordingly, then hit 'Start evaluating' button (Fig. 8(a)).
 III. View the result carefully. If you have questions about entries on the page, you may click on the question mark in a blue box after each entry (Fig. 8(b)).

> *Note*: The lower the DOPE score is, the better the model.
> *Note*: Please remember to save result web page for future reference.

7.   Viewing and rendering the protein model with PyMol

After modeling is finished, you will get a pretty figure of your protein model. However, if you wish to see it in another angle, zoom to the important sites. Here we introduce PyMol,[16] which is capable of displaying and rendering your protein model as you wish.

Materials

• The model(s) created by any of the protein modeling software.

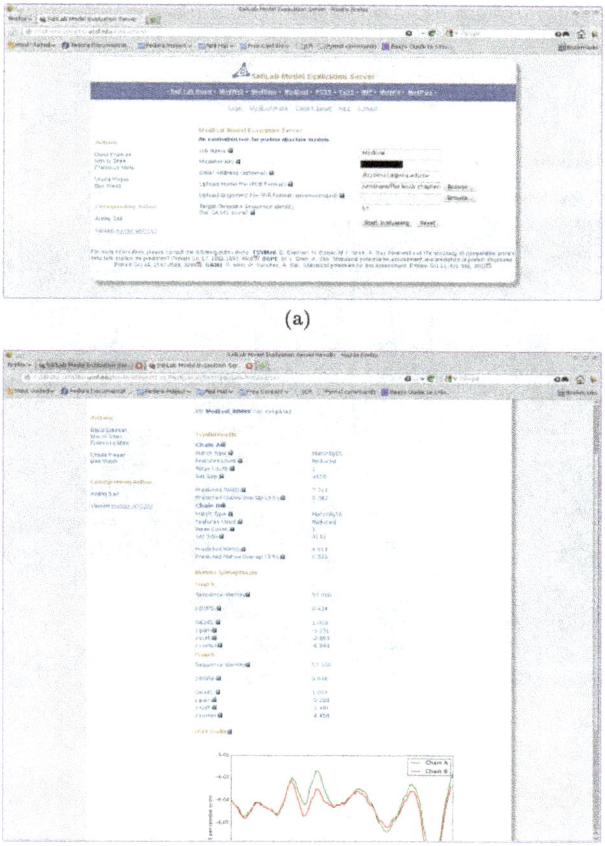

(a)

(b)

Figure 8. Evaluate model with ModEval. (a) Job submission form and (b) result page, containing Dope score, GA341 score, and DOPE profile (the figure in the lower part of the page) of the protein.

- PyMol software. Please install it before continuing reading. For Microsoft Windows users, you may find pre-compiled pymol binaries in http://www.lfd.uci.edu/~gohlke/pythonlibs/#pymol or search for "precompiled binary Windows" in your search engine. For Linux users, you may consult your software repositories, or PyMol official site.

Procedure

I. Launch PyMol.

After launch, you will see a window for 3D model view and another window with menu bar, lots of buttons and a large text area (Fig. 9). Both windows are capable of receiving text commands.

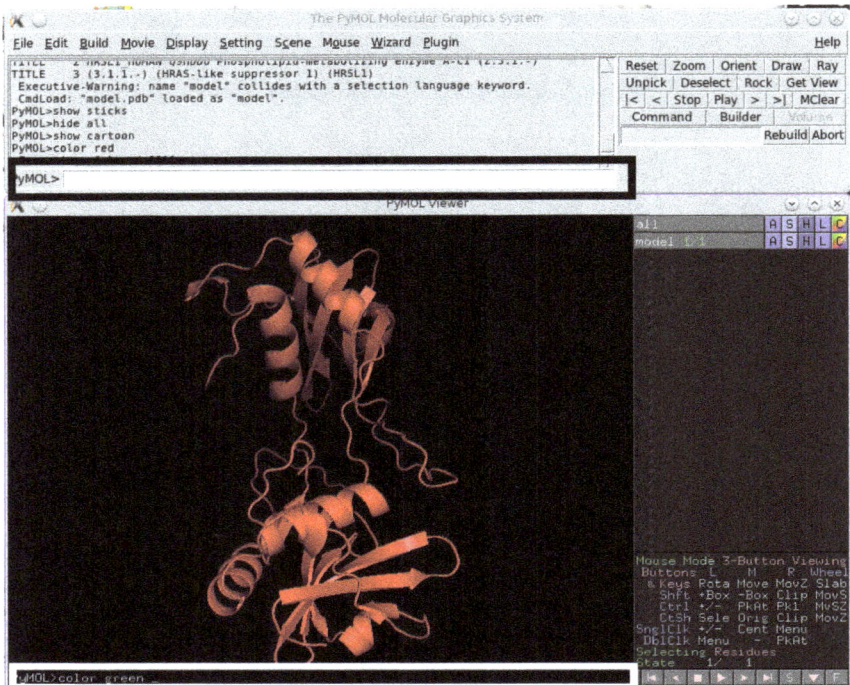

Figure 9.   Screenshot of PyMol application. Black box and white box indicates the place where you can input PyMol commands.

II. Load your structure file by 'load' command.
    load myProtein.pdb, myProtein
    You may use 'pwd' command to see where current directory is. You may want to use 'cd' command to switch into the directory where your structure files are stored, then use 'load' command.
    You might also launch and open your structure file by double-clicking on the structure file in your file manager if the installer supported it.

III. Use the mouse to adjust pose of the model until you feel satisfied.
    Drag with mouse left button pressed to rotate, drag with mouse right button pressed to zoom, and drag with mouse middle button to pan the view.

IV. You may adjust display style of your model with other commands. To see all possible commands, please go to http://www.pymolwiki.org/index.php/Category:Commands, or search 'pymol wiki command' in your search engine.
    You may also find other useful resources there, in PymolWiki (http://www.pymolwiki.org/)

V. Use 'png' command to save your figures:

`png molecular_image.png width=640 height=480 ray=1`

If you want to preserve current poses for possible uses in the future, please use 'save' command to save PyMol session stat and use 'load' command to load it when needed.

`save pymolSession.pse`

`load pymolSession.pse`

References

1. Harikishore A, Li E, Lee J *et al.* (2015) Combination of pharmacophore hypothesis and molecular docking to identify novel inhibitors of HCV NS5B polymerase. *Mol. Divers.* **19**(3): 529–539.
2. Kang HJ, Lee C, and Drew D. (2013) Breaking the barriers in membrane protein crystallography. *Int. J. Biochem. Cell Biol.* **45**(3): 636–644.
3. Chowdhury S, Lee MC, Xiong G *et al.* (2003) *Ab initio* folding simulation of the Trp-cage mini-protein approaches NMR resolution. *J. Mol. Biol.* **327**(3): 711–717.
4. Levinthal C. (1968) Are There Pathways For Protein Folding? *Extrait du Journal de Chimie Physique.* **65**(1): 44–45.
5. van Gunsteren WF and Berendsen HJC. (1990) Computer Simulation of Molecular Dynamics: Methodology, Applications, and Perspectives in Chemistry. *Angew. Chem. Int. Ed.* **29**(9): 992–1023.
6. Chothia C and Lesk AM. (1986) The relation between the divergence of sequence and structure in proteins. *EMBO J.* **54**: 823–826.
7. UniProt Consortium, (2015) UniProt: A hub for protein information. *Nucleic Acids Res.* **43**(Database issue): D204–D212.
8. Pruitt KD, Tatusova T, Brown GR *et al.* (2012) NCBI Reference Sequences (RefSeq): Current status, new features and genome annotation policy. *Nucleic Acids Res.* **40**(Database issue): D130–D135.
9. Biasini M, Bienert S, Waterhouse A *et al.* (2014) SWISS-MODEL: Modelling protein tertiary and quaternary structure using evolutionary information. *Nucleic Acids Res.* **42**(Web Server issue): W252–W258.
10. Yang J, Yan R, Roy A *et al.* (2015) The I-TASSER Suite: Protein structure and function prediction. *Nat. Methods.* **12**(1): 7–8.
11. Eswar N, John B, Mirkovic N *et al.* (2003) Tools for comparative protein structure modeling and analysis. *Nucleic Acids Res.* **31**(13): 3375–3380.
12. Pieper U, Webb BM, Dong GQ *et al.* (2014) ModBase, a database of annotated comparative protein structure models and associated resources. *Nucleic Acids Res.* **42**(Database issue): D336–D346.
13. Haas J, Roth S, Arnold K *et al.* (2013) The Protein Model Portal — a comprehensive resource for protein structure and model information. *Database: The Journal of Biological Databases and Curation 2013: bat031.*
14. Melo F, Sanchez R, and Sali A. (2002) Statistical potentials for fold assessment. *Protein Science: A Publication of the Protein Society.* **11**(2): 430–448.
15. Benkert P, Kunzli M, and Schwede T. (2009) QMEAN server for protein model quality estimation. *Nucleic Acids Res.* **37**(Web server issue): W510–W514.
16. Schrodinger LLC, The PyMOL Molecular Graphics System, Version 1.3r1. In 2010.

11. Docking Simulation

Chia-Hsien Lee and Hsueh-Fen Juan*

*Graduate Institute of Biomedical Electronics and Bioinformatics,
National Taiwan University, Taipei, Taiwan
yukijuan@ntu.edu.tw

1. Introduction

In computational drug discovery, there are two strategies to do drug design: ligand-based and structure-based. In ligand-based drug design, a set of common ligand of target protein is used to deduce, from the similarity of that set of ligands, a novel drug. In structure-based drug design, information of protein structure and putative compounds are utilized in the drug discovery process. For example, in the work of Bessa et al.,[1] docking simulation was performed to explore the reason of synergy interaction between the two specific compounds and antibiotic oxacillin. Li et al.[2] used docking simulation (structure-based) and pharmacophore (ligand-based drug design) and then found two putative lead compounds for inosine 5'-monophosphate dehydrogenase, which is an attractive target in immunosuppressive, anti-cancer, antiviral, and antiparasitic therapeutic strategies. We will focus on structure-based drug design, especially docking simulation, from now on.

Docking simulation is the most widely used technique in structure-based drug design. The structures of the target protein and compound are input into a docking algorithm which will find the best pose of their binding. Since it is all about finding the "best" poses of the binding of protein and the compound, docking simulation is essentially an optimization problem, and thus can be solved by optimization algorithms after proper formulation. Most docking simulation software packages use various optimization algorithms for their calculation jobs.

Autodock Vina,[3] a standalone docking and an open-source application package, is widely used when a biologist needs to run docking simulation.

In essence, it accepts a protein structure file, a compound structure file, and some other options (such as binding site location) as input, running docking simulation with an optimization algorithm called Iterated Local Search global optimizer,[4] then produce a list of binding poses and corresponding score. However, due to its design, one should use a controller script or application to input ligand-protein target pairs to Autodock Vina for virtual screening.

In this chapter, we will introduce the process of running docking (Fig. 1) from scratch and guide you through the whole process, including fetching and preparing those structure files and databases, feeding them into Autodock Vina. Additionally, we will also guide you through the using of SwissDock,[5] a fully automatic docking system.

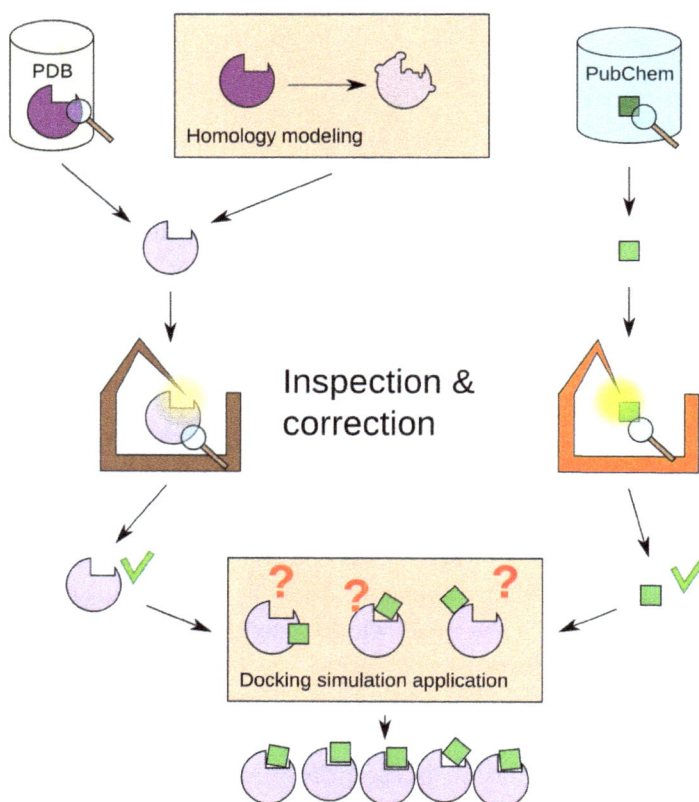

Figure 1. Overall flowchart of docking simulation. First, structure of the ligand and the protein target should be fetched from the respected databases. Second, the structures should be inspected and modified, if necessary, to be able to be used by docking simulation engine. After the preparations are done, the docking pose(s) can be generated by the docking simulation application.

2. Get ligand structures from PubChem

Here we'll guide you to query PubChem[6,7] and download your query result into a ligand database (Fig. 2).

One should take note that downloading everything from PubChem is possible but not recommended since there are huge number of compounds stored in PubChem database.

Materials

- Information of your ligands. For example, you should know the name of the ligand, to which they are similar, or molecular weight range and so on.

(a)

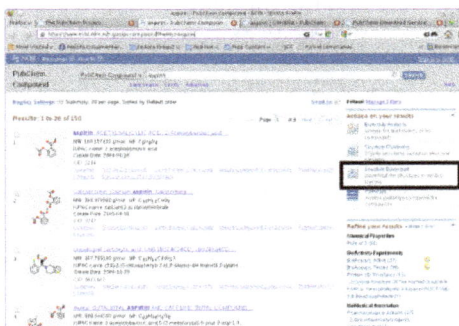

(b)

(c)

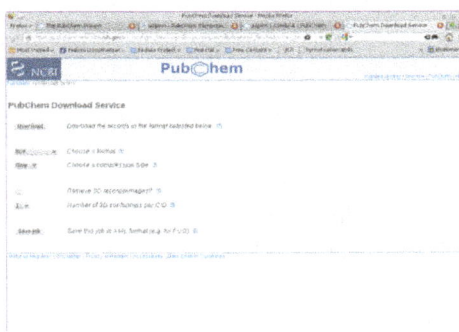

(d)

Figure 2. Searching ligands from PubChem. (a) PubChem entrance interface. (b) PubChem search result page and the link bringing the users to the batch download interface (black box). (c) Detailed information page of single compound. It may be downloaded with the download link provided (black box). (d) Batch download interface of query result.

Procedure

I. Navigate to PubChem by using its URL (https://pubchem.ncbi.
   nlm.nih.gov/) or by querying the search engine (e.g., Google).

II. Do search according to the ligand information you have (Fig. 2(a)).
   If you have some keywords about the ligand, do searching directly in
   the search box and click the "Go" button.
   If you have more information about the ligand (e.g., molecular weight
   range, hydrogen bond acceptor count, etc), go to advanced search page
   by clicking the "advanced" link on the right side of the "Go" button,
   fill the blanks and hit "Search" button in the advanced search page.
   If you have the ligand structure, click "structure search" on the right
   panel, go to structure search page, fill your structure and select the
   searching method (e.g., similarity search, identity search, substructure
   search, or superstructure search), then hit "Search" button.

III. No matter which way you choose to do your query, you will be taken to
   the result page. View the results carefully (Fig. 2(b)).
   If your goal is a specific compound, please go through the list one by one
   by going into the detail page. You may download the structure by using
   "Download → 2D structure" or "Download → 3D conformers" in the
   detailed page. After you finish your download, take note of PubChem
   CID of this compound for future reference (Fig. 2(c)).
   If you want to grab all compounds that fit into your query, please find
   "Structure Download" button under "Actions on your results" section
   in the search result list. Saving in SDF format is recommended. Don't
   forget to take note of your query date (Fig. 2(d)).

3.   Get ligand structures from ZINC

Here we will guide you to query ZINC[8] and download your query result into
a ligand database (Fig. 3). Please note that you may download only up to
1000 molecules at once by using this method.

Materials

• Information of your ligands. For example, you should know the name
  of the ligand, to which they are similar, or molecular weight range
  and so on.

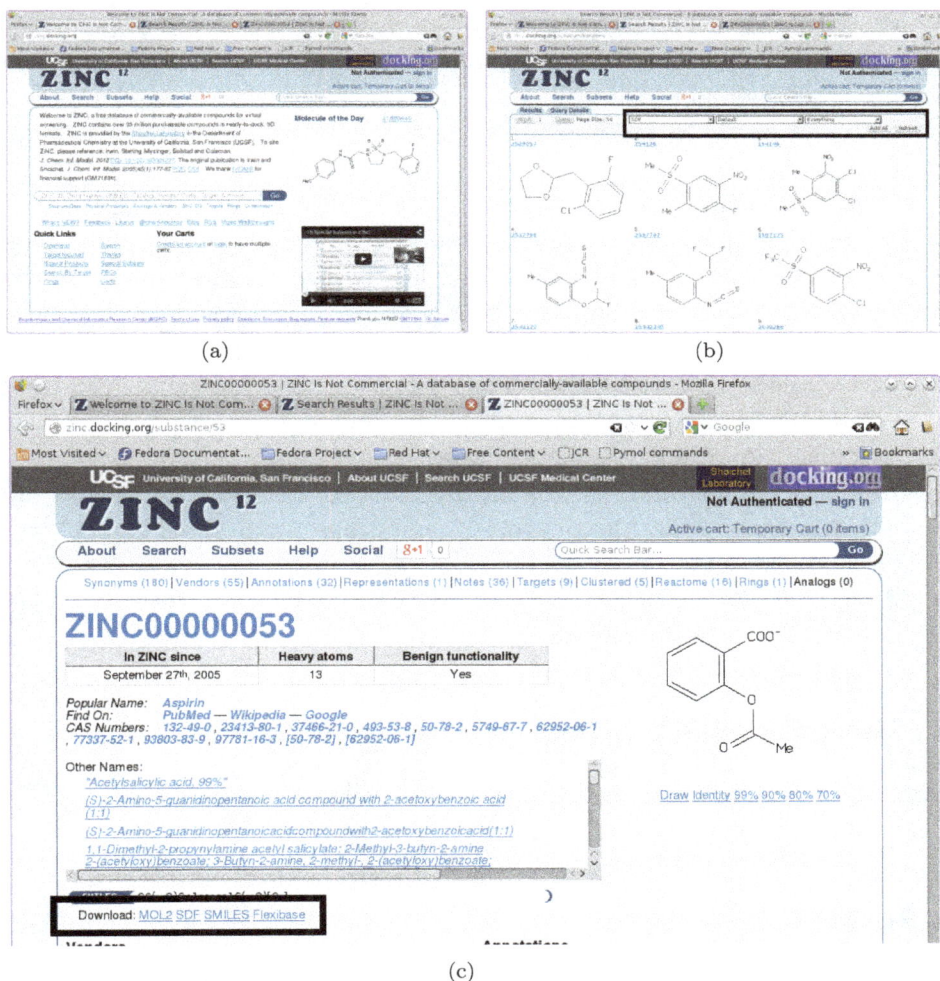

(a)

(b)

(c)

Figure 3. Searching ligands from ZINC database. (a) ZINC entrance interface. (b) ZINC search result page and the control interface for batch downloading (black box). (c) Detailed information page of single compound. Structure of this compound can be downloaded with the links indicated by the black box.

Procedure

I. Navigate to official site of ZINC database by using its url (http://zinc.docking.org/), or search "ZINC database" in your search engine.

II. You can do quick search by using the quick search bar in the upper right side of the page. You may move your mouse onto "Search" button in the bar under the large "ZINC" logo. A small panel will pop up and showing all possible search methods. Choose the method which fit your needs most (Fig. 3(a)).

III. Irrespective of the method used, query results will be listed on screen (Fig. 3(b)).

If you are using quick search and there is only one result (e.g., search "aspirin" in quick bar), you will be brought into compound detail page describing that chemical. In this case, just download its structure in this detail page. (Download options are under the SMILES string) (Fig. 3(c))

IV. Under the bar containing "Search", "Subsets", "Help" buttons you can find a combo box with "Overview" text on it. It is the left of the three combo boxes. Click on that combo box, browse to "Download" section and pick the format you like (Fig. 3(b)).

If you wish to filter the result by their purchasability, make your adjustment by using the rightmost combo box (with "Everything" text on it by default).

V. After you have done, click the "Refresh" button on the right. The download will begin soon (Fig. 3(b)).

4. Get pre-defined ligand database from ZINC

Here, we will guide you to fetch pre-defined ligand database from ZINC (Fig. 4).

Materials

- If you are using Microsoft Windows, obtain `wget` downloader and place them into a place where it can be located by the operation system. (i.e., place `wget` into any directory indicated in ``%PATH%`` environment variable)

Procedure

I. Navigate to official site of ZINC database by using its url (http://zinc.docking.org/), or search "ZINC database" in your search engine.

II. Move your mouse to "Subsets" button in the bar under the large "ZINC" logo. You will see a small panel popped up. On that panel, you can see "By catalog property target ring cart" (four options). You may click in them and explore to find the subset you need (Figs. 4(a)–(b)).

(a)   (b)

(c)

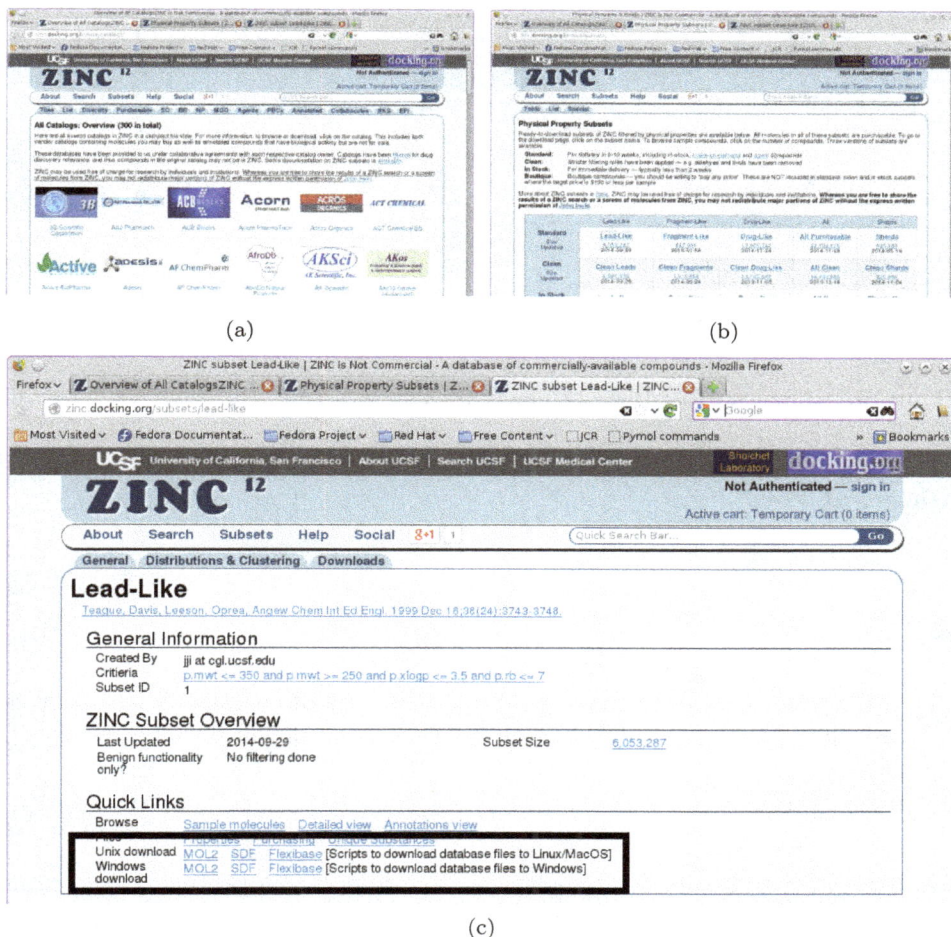

Figure 4. Subset listing of ZINC database (a) by catalog, (b) by property and (c) subset information page. Black box: script for subset database downloading.

III. After finding the subset which fits your needs, you may download them according to the instruction on the page (Fig. 4(c)). To download the subset, use the download script provided in the page.

We recommended downloading those subsets in SDF or MOL2 format.

5. Get ligand database from DrugBank

Here we will guide you to fetch pre-defined ligand database from DrugBank (Fig. 5).[9–12]

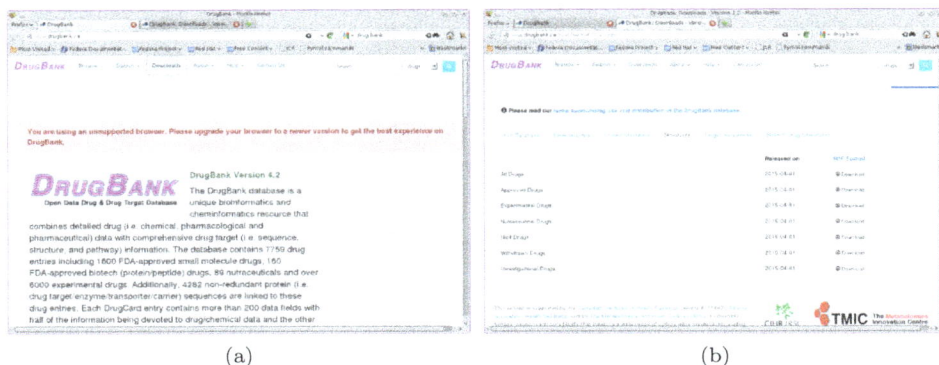

(a)   (b)

Figure 5. Download DrugBank database. (a) Entrance page. (b) Database download page.

Materials

- (None)

Procedure

I. Open your browser and navigate to official site of DrugBank by using its url (http://www.drugbank.ca/), or search "DrugBank" in your search engine.

II. Find and click on the "Download" button on the top of the website screen (Fig. 5(a)).

III. Now you are in the download page. You may find a tabbed panel under the text "Please read our terms surrounding use and distribution of the DrugBank database". Currently you are in "Full database" tab. Please switch to "Structures" tab.

IV. You may find a list of available subsets for download. Please select the one suitable for your needs (Fig. 5(b)).
    We recommend you to use Approved Drugs set if you want to find new use of those old but approved drugs.

6. Running docking simulation with Autodock Vina

In this section, we will tell you how to run docking simulation with Autodock Vina. Since Autodock Vina operates in command-line mode (i.e., you should use command prompt in Microsoft Windows or terminal applications such as Konsole or GNOME terminal in GNU/Linux), you should possess some knowledge of basic file-related operations under text mode.

For Microsoft Windows users, installing MSYS system (from http://www.mingw.org/wiki/msys), Cygwin (https://www.cygwin.com/), or similar systems are recommended.

The approach described here is different from the one in the official site. If you prefer the official one, please navigate to http://vina.scripps.edu/tutorial.html or search 'vina video tutorial' in your search engine.

Materials

- Ligand structure file which contains only one compound inside.
- Protein target structure file.
- OpenBabel[13] installation.
- Autodock Vina application which can be downloaded from http://vina.scripps.edu/download.html.

Procedure

I. Navigate into an appropriate directory (e.g., where your ligand and protein structure files are stored, or where you want to store your result) in the text-based shell of your choice.

   In Windows, use command prompt; in GNU/Linux, use of any virtual terminal application will do.

II. Use the following command to prepare ligand in your shell prompt:

```
cat ligand.sdf |obabel -isdf --AddPolarH -opdbqt >
ligand.pdbqt
```

or

```
obabel -isdf ligand.sdf --AddPolarH -opdbqt Oligand.pdbqt
```

If you are using ligand in mol2 format, change underlined part to mol2; if you are using ligands in pdb format, change underlined part to pdb. If the coordinates of the ligand is in 2D, please add --gen3D switch, like this:

```
cat ligand.sdf | obabel -isdf --AddPolarH --gen3D -opdbqt
> ligand.pdbqt
```

or

```
obabel -isdf ligand.sdf --AddPolarH --gen3D -opdbqt
-Oligand.pdbqt
```

III. Load the protein structure file into a molecular editor (e.g., PyMol) and remove any water molecules in it. Remove any other things you don't need in this process.

Do not close your molecular editor yet! We need it in the next step.

To accomplish such task with PyMol, you may need the following commands: `select`, `remove`, and `save`.

For all possible commands usable in PyMol, along with detailed description about how to use those commands, please navigate to PyMol wiki here: http://www.pymolwiki.org/index.php/Category: Commands.

IV. Check the dimension of binding site box of your protein in your molecular editor and take note about it.

In essence, you should take note of **center** and **size** of the binding site box.

In PyMol, you may use `select` command to pick the native ligand (i.e., the ligand comes with the protein crystal) or amino acid residues near the binding site, then use `get_extent` command to get dimensions of this box.

V. Use the following command to prepare protein structure in your shell prompt:

```
cat protein.pdb |obabel -ipdb --AddPolarH -xr -xc
-opdbqt > protein_o.pdbqt
```

or

```
obabel -ipdb protein.pdb --AddPolarH -xr -xc -opdbqt
-Oprotein_o.pdbqt
```

VI. Edit produced `protein_o.pdbqt` file, then save as `protein.pdbqt`: Retain lines starting with `ATOM` tag and remove all others.

You may use application `grep` for this task if you are using GNU/Linux, Cygwin or MinGW:

```
grep ATOM protein_o.pdbqt > protein.pdbqt
```

VII. Write the configuration file for docking like the following:

```
receptor = protein.pdbqt
ligand = ligand.pdbqt
out = output.pdbqt
center_x = 2.5
center_y = 2.5
```

```
center_z = −2.5
size_x = 10
size_y = 10
size_z = 10
energy_range = 4
exhaustiveness = 16
```

Modify the fields when needed.

VIII. Run Autodock Vina by (assume the configuration file is saved to myConfig.conf)

```
vina --config myConfig.conf
```

Docking will be finished soon if **exhaustiveness** is not set too high.

IX. The result will be stored in **output.pdbqt** or where you specify in 'out' field of the configuration file. Please open it with your text editor.

X. After browsing the whole file, you may note that **vina** will give some (usually more than one) poses, and sorted by binding energy from low to high. (Since higher binding energy means that the binding mode is not stable, thus may not be preferred).

From the result file, you can see a line like the following:

REMARK VINA RESULT: −6.6   0.000   0.000

The number marked with underline (−6.6 here) is binding energy. In most cases, you can just use the first model (pose) in the result pdbqt file without caring about others, since it will be the one with lowest binding energy.

> The lower the binding energy is, the better the pose is.

XI. If you want to view the result in your molecular editor, please load the structure and the result file into the same view.

In PyMol, use the following commands to view the binding:

```
load protein.pdbqt, target
load result.pdbqt, ligands
```

7. Running docking simulation with SwissDock (Fig. 6)

SwissDock is a docking web service provided by Swiss Institute of Bioinformatics. It can automatically setup input protein and ligand for the user.

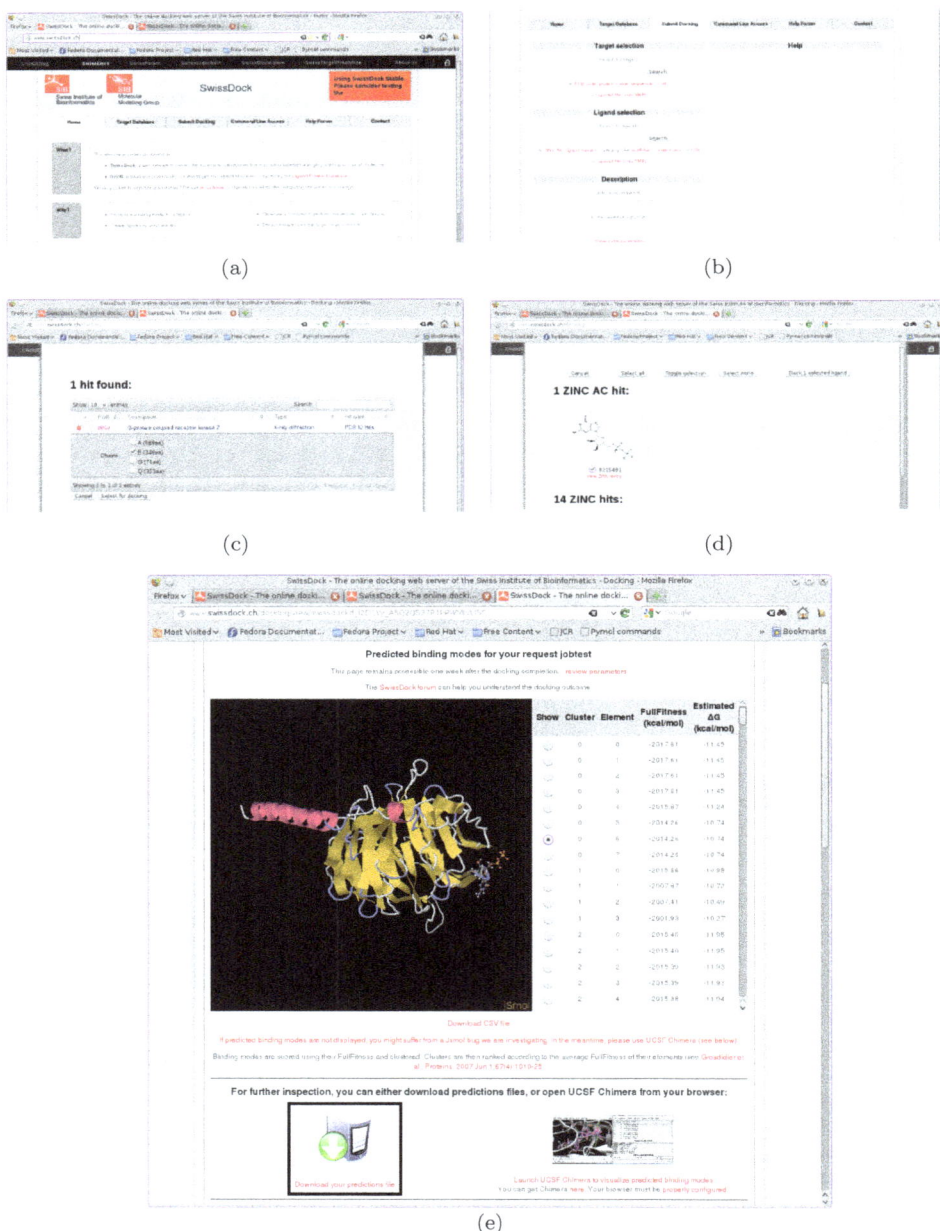

(a)

(b)

(c)

(d)

(e)

Figure 6.   Docking with SwissDock. (a) Entrance page of SwissDock. (b) Job submission page. (c) Search target by PDB ID. (d) Search ligand by ZINC ID. (e) Result page. In this page, you can preview selected docking poses listed in the right panel in the JavaScript-based viewer on the left panel. To download the result for further analysis, use the download button in the bottom of this screen (black box).

Materials

- Ligand file (containing only single compound)
- Protein target structure file

Procedure

I. In your browser, navigate to http://www.swissdock.ch/, or search for "SwissDock" in your search engine.

II. You're in "Home" tab right now. Please switch to "Submit Docking" tab (Fig. 6(a)).

III. After switching to "Submit Docking" tab, fill the fields in target selection, ligand selection, and description panels (Fig. 6(b)).

You may submit your file by clicking on "upload file (max 5 MB)". You may also submit by PDB ID (target) or ZINC ID (ligand), etc. if you already know the ID or name of the protein target or ligand (Figs. 6(c)–6(d)).

After you see "Successful setup" text appear right after you submit your target or ligand, you can click on "inspect" on the right of "Successful setup" to review your model before running docking simulation. It is highly recommended that you review your model before doing the computation.

It is highly recommended that you fill your e-mail such that you will receive a notification after the job is finished.

You can assign binding site (if you have such information) and set docking operation type after clicking on "Show extra parameters" above "Start docking" button.

IV. It takes about 30 minutes to finish a docking run.

If you did not receive any notification about your docking, you may have a typo in your email address and should try again later. If such condition persists, please contact maintenance team by using "contact" tab.

V. You may view the result online if your computer system has java support, or you may view the result using UCSD chimera by following the instructions in the result page, or you may download the result into your hard disk and view them with your favorite molecular viewer (Fig. 6(e)). The result poses are grouped into clusters, and the clusters are sorted by average full fitness. In general, element 0 of cluster 0 has the best full fitness.

The smaller the full fitness is, the better the predicted binding pose.

It's highly recommended that you download and save the result and save them in your hard drive.

In the downloaded package, information about docking fitness is recorded in `REMARK` part of each pose of `clusters.dock4.pdb`. Pre-processed structure of target protein is in `target.pdb`.

To view using PyMol, you need `target.pdb` and `clusters.dock4.pdb` in the downloaded package. To separate `clusters.dock4.pdb` into poses, you need to use `TER` as separator. In most cases, the best pose is in the first part of `clusters.dock4.pdb`, thus you can get the best pose by copying all the text before the first `TER` and paste them into a new file.

References

1. Bessa LJ, Palmeira A, Gomes AS, Vasconcelos V, Sousa E, Pinto M, and Martins da Costa, P. (2015) Synergistic effects between thioxanthones and oxacillin against methicillin-resistant staphylococcus aureus. *Microb. Drug. Resist.* **21**(4): 404–415.
2. Li RJ, Wang YL, Wang QH *et al.* (2015) Silico design of human IMPDH inhibitors using pharmacophore mapping and molecular docking approaches. *Computational and Mathematical Methods in Medicine* 418767.
3. Trott O and Olson AJ. (2010) AutoDock Vina: Improving the speed and accuracy of docking with a new scoring function, efficient optimization, and multithreading. *J. Comput. Chem.* **31**(2): 455–461.
4. Baxter, J. (1981) Local optima avoidance in depot location. *J. Oper. Res. Soc.* **32**(9): 815–819.
5. Grosdidier A, Zoete V, and Michielin O. (2011) SwissDock, a protein-small molecule docking web service based on EADock DSS. *Nucleic Acids Res.* **39** (Web Server issue): W270–W277.
6. Bolton EE, Wang Y, Thiessen PA *et al.* (2008) Chapter 12 - PubChem: Integrated platform of small molecules and biological activities. In *Annual Reports in Computational Chemistry*, Ralph, A. W.; David, C. S., Eds. Elsevier: Volume 4: 217–241.
7. Wang Y, Xiao J, Suzek TO *et al.* (2009) PubChem: A public information system for analyzing bioactivities of small molecules. *Nucleic Acids Res.* **37** (Web Server issue): W623–W633.
8. Irwin JJ, Sterling T, Mysinger MM *et al.* (2012) A free tool to discover chemistry for biology. *J. Chem. Inf. Model.* **52**(7): 1757–1768.
9. Law V, Knox C, Djoumbou Y. *et al.* (2014) DrugBank 4.0: Shedding new light on drug metabolism. *Nucleic Acids Res.* **42** (Database issue): D1091–D1097.
10. Knox C, Law V, Jewison T *et al.* (2011) DrugBank 3.0: A comprehensive resource for 'omics' research on drugs. *Nucleic Acids Res.* **39** (Database issue): D1035–D1041.
11. Wishart DS, Knox C, Guo AC *et al.* (2008) DrugBank: A knowledgebase for drugs, drug actions and drug targets. *Nucleic Acids Res.* **36** (Database issue): D901–D906.

12. Wishart DS, Knox C, Guo AC *et al.* (2006) DrugBank: A comprehensive resource for *in silico* drug discovery and exploration. *Nucleic Acids Res.* **34** (Database issue): D668–D672.

13. O'Boyle N, Banck M, James C *et al.* (2011) Open babel: An open chemical toolbox. *J. Cheminform.* **3**(1): 1–14.

Index

acute lymphoblastic leukemia (ALL), 95
Akaike information criterion (AIC), 109
ALL1–AF4 rearrangement, 95
Autodock Vina, 136

bioinformatics, 1
biological modeling, 103
biological networks, 103
biomarker, 4

computer-aided drug discovery, 113
Cytoscape, 64

Database for Annotation, Visualization
    and Integrated Discovery (DAVID),
    63
differential equation, 104
differentially expressed (DE) genes, 95
dimethyl labeling, 40
discrete dynamic models, 104
docking simulation, 129
drug discovery, 4
DrugBank, 135
dynamic modeling, 4
DynaPho, 76

EnrichmentMap, 63

function enrichment analysis, 80
functional enrichment analysis, 63

galaxy, 50
gel-assisted digestion and gel extraction,
    25
gene expression, 92
gene ontology, 63
Gene Set Enrichment Analysis (GSEA),
    63

gene–gene correlation networks, 99
genome, 2

homology modeling, 114
hydroxyl acid-modified metal oxide
    chromatography (HAMMOC), 40
hypergeometric test, 97

I-tasser, 114
Illumina, 12
isobaric tags for relative and absolute
    quantitation (iTRAQ), 25
isotope-based methods, 25
isotope-coded affinity tag (ICAT), 25
iTRAQ, 65

kinase activity profile, 84
kinase/phosphatase–substrate association,
    87
Kyoto Encyclopedia of Genes and
    Genomes (KEGG) pathways, 97

mass spectrometry, 63
mathematical models, 103
maximum likelihood, 107
metabolic pathways, 63
metabolome, 4
ModBase, 114
model selection, 109
ModEval, 114
ModWeb, 114
molecular mechanism, 5
motif enrichment analysis, 80

nanoLC-MS/MS, 40
network biology, 4
next-generation sequencing (NGS), 11, 49

omic data, 91
omics, 1

parameter identification, 107
Pearson correlation coefficients (PCCs),
    99
phase transfer surfactant-aided trypsin
    digestion, 40
Philadelphia chromosome, 95
phosphoproteomics, 75
post-translational modifications, 63
profile clustering, 78
Protein Model Portal, 114
protein phosphorylation, 39
protein–protein interaction network, 63,
    82
proteome, 3
PubChem, 131

Q–Q plot, 96
Qmean server, 114

R programming language, 91
reporter ions, 25
RNA-seq, 49

shotgun proteomics, 23, 63
signaling pathways, 63
Significant Analysis of Microarray (SAM),
    95
Stable isotope labeling by amino acids in
    cell culture (SILAC), 25
stepwise regression method, 110
structure-based drug design, 129
SwissDock, 139
SwissModel, 114
systems biology, 1

tandem mass spectrometry (MS/MS), 23
target identification, 4
titanium dioxide (TiO_2) enrichment, 39
transcriptional regulations, 63
transcriptome, 3, 11
TruSeq, 12
TXT file, 64

ZINC, 132

www.ingramcontent.com/pod-product-compliance
Lightning Source LLC
Chambersburg PA
CBHW081519190326
41458CB00015B/5402